ADHESIVES TECHNOLOGY COMPENDIUM 2017

adhesion ADHESIVES & SEALANTS

Industrieverband Klebstoffe e. V.

(German Adhesives Association)

Springer Vieweg

Editor: adhesion ADHESIVES & SEALANTS
Abraham-Lincoln-Straße 46
D-65189 Wiesbaden
Phone +49 (0) 6 11-78 78-2 83
Fax +49 (0) 611-78 78-4 95
www.adhaesion.com
Email: adhaesion@springer.com

Supported by:
Industrieverband Klebstoffe e. V.
German Adhesives Association
Völklinger Straße 4 (RWI-Haus)
D-40219 Düsseldorf
Phone +49 (0) 2 11-6 79 31 10
Fax +49 (0) 2 11-6 79 31 33

Publishing house: Springer Vieweg | Springer Fachmedien Wiesbaden GmbH
Abraham-Lincoln-Straße 46
D-65189 Wiesbaden
www.springer-vieweg.de

| Austria (A) | Swiss (CH) | Germany (D) | Netherland (NL) |

Layout: satzwerk mediengestaltung · D-63303 Dreieich

ISBN 978-3-658-18741-5

Nominal sum: € 25.90

Dear Reader,

Adhesive bonding is without a doubt a key technology of the 21st century. Today, there is hardly any industry or business segment that does not rely on the use of this innovative, reliable joining method. Adhesive bonding is indispensable when it comes to joining different materials while preserving their characteristics and offering long-term stability. The opportunities for using new, reliable construction methods can only be exploited in conjunction with innovative adhesive systems. In addition to the joining process itself, other characteristics can also be integrated into bonded components, for instance by balancing the differing dynamics of joined parts, providing corrosion protection or vibration absorption or sealing against liquids and gases. More than any other joining technology, adhesive bonding allows sophisticated designs to be created, because it offers the best possible combination of technological sophistication, cost-effectiveness and a low environmental impact. Adhesive bonding technology is utilised without exception by all sectors of industry.

The *Industrieverband Klebstoffe e.V.* (German Adhesives Association, IVK) represents the technical and economic interests of the German adhesives industry. It was founded in 1946 and is now the world's largest and, in terms of its comprehensive portfolio of services, the world's leading association for adhesive bonding technology.

In this compendium the German Adhesives Association, together with its sister associations – the Association of the Adhesives Industry in Switzerland (FKS), the Austrian Association for Flooring Adhesives (FCIO) and the Association of the Dutch Adhesives and Sealants Industry (VLK) – gives an insight into the world of the adhesives industry.

One of the main tasks of these associations is to provide regular information about adhesive bonding as a key technology, as well as about manufacturers of innovative adhesive systems and the activities of the industry's organisations. This compendium contains important facts about the adhesive industry and its associations, as well as describing the extensive product and service profiles of adhesives manufacturers, key system partners and scientific institutes. Together with the editorial staff of "adhäsion KLEBEN & DICHTEN", we are pleased to present the 15th edition of our Adhesives Technology Compendium.

Dr. Boris Tasche
President of
Industrieverband Klebstoffe e.V.

Ansgar van Halteren
Senior Executive of
Industrieverband Klebstoffe e.V.

German
Adhesives
Association

Industrieverband Klebstoffe e.V.

adhesion ADHESIVES & SEALANTS

MADE IN GERMANY

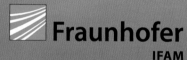

Fraunhofer
IFAM

EMICODE® GEV

Industrieverband Klebstoffe e. V.
Völklinger Straße 4 (RWI-Haus)
D-40219 Düsseldorf
Phone +49 (0) 2 11-6 79 31 10, fax +49 (0) 2 11-6 79 31 33

Industrieverband
Klebstoffe e.V.

List of Advertisers

Cover: Optimization of adhesive application (© Fraunhofer IFAM, Bremen, Germany)

DIN 2304

Professional Implementation of Adhesive Bonding Processes

The DIN 2304 standard, which came into effect in March 2016, established the mandatory state-of-the-art for the professional implementation of adhesive bonding processes.

According to DIN EN ISO 9001, adhesive bonding is, just as welding, defined as a special process. This is a process whose results cannot be fully examined non-destructively. A sustainable and reproducible adhesive bonding process in accordance with DIN EN ISO 9001 means developing and pursuing a strict strategy failure prevention. This can be achieved by detailed specification of a bonding process in a quality management system customized for the particular application.

The new DIN 2304 describes the professional implementation of adhesive bonding processes and applies to all bonds whose primary function is to transmit mechanical loads. It is valid for all branches of industry that do not have standardized adhesive bonding applications. This results in a very wide user field from e.g. the following industrial sectors:

- Automotive industry
- Aircraft manufacture
- Plant engineering
- Electrical construction
- Microsystem technology and optics
- Construction industry
- Shipbuilding
- Special vehicle construction
- Packaging industry
- Tool manufacture
- Medical technology
- Textile industry

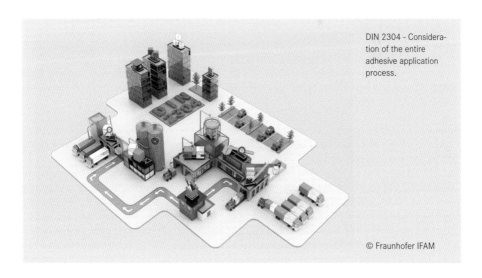

DIN 2304 - Consideration of the entire adhesive application process.

© Fraunhofer IFAM

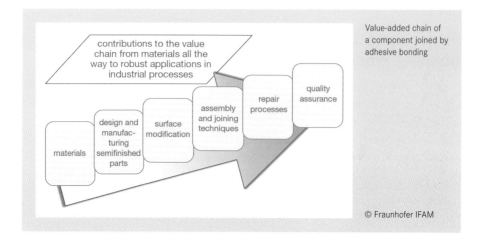

Value-added chain of a component joined by adhesive bonding

contributions to the value chain from materials all the way to robust applications in industrial processes

materials | design and manufacturing semifinished parts | surface modification | assembly and joining techniques | repair processes | quality assurance

© Fraunhofer IFAM

The DIN 2304 standard presents new challenges to all industries that use adhesive bonding technology. First, all adhesive joints must be categorized into safety classes, because these determine the quality requirements for the entire adhesive bonding process. The safety classes are based on the risks that exist in the event of a failure of the adhesive bond. Class S1 "high safety requirements" has the highest level of effort and expense for quality assurance. If conversely an adhesive bond is classified as class S4 "no safety requirements", then no special requirements are necessary.

From planning to repair

If the adhesive bonds used in a company's operations are classified in safety classes S1 to S3, then the challenge in this instance is to develop the right concept for the business as well as the quality management system to ensure the aspects related to adhesive bonding technology. This starts with planning which materials are to be used and continues all the way to potential product repair.

When creating the concept, all areas of the business that could influence the product which is to be bonded later must be taken into account. This includes areas where this influence is not immediately visible.

Considering the entire process chain of the adhesive bonding manufacturing, this starts by examining the contracts made with the customer. This is to ensure that all of the adhesive bonding requirements can be implemented. If the adhesive bonds are not manufactured by the company, but rather come from suppliers, the subcontract must contain a clear definition of the adhesive bonding requirements. The development process is important in this case. For example in construction, the loads acting on the bond must be ensured to be less

Optimization of adhesive
application

© Fraunhofer IFAM

than the load carried by the joint (load-bearing capacity). During the process planning, the process window and tools necessary to produce the adhesive bonds should be identified and work instructions should be created.

For the production phase, the DIN 2304 introduces requirements regarding company infra-structure, adhesive bonding work areas for production and repair, storage and logistics as well as the quality assurance measures that accompany manufacturing. As already intro-duced in the DIN EN ISO 9001, there are also requirements regarding the monitoring of measuring, testing and manufacturing resources.

Accredited training

An important aspect of the entire adhesive bonding process is the training of staff in adhe-sive bonding technology. Contrary to the approach that only allows adhesives for certain applications; the DIN 2304 is based on the professional execution of tasks such as adhesive selection, adhesive qualification and the manufacturing of the adhesive bonds.

For further training in adhesive bonding technology, DIN 2304 uses the long-standing DVS / EWF personnel qualifications for European Adhesive Bonder (EAB), European Adhesive Specialist (EAS) or the European Adhesive Engineer (EAE). These courses are offered in accredited training centers in both German and English as well as in-house courses which are also translated into the local language.

For the initial introduction of the DIN 2304, it is firstly recommended that the current status of the adhesive bonding process in its entirety while in operation is observed. The following steps should be carried out:

Adhesive bonding
technical training at
Fraunhofer IFAM

© Fraunhofer IFAM

1. Analysis of the entire adhesive bonding process for an existing or planned production including, where appropriate, the implementation of a Failure Modes and Effects Analysis (FMEA) of the adhesive bonding aspects.
2. Evaluation of compliance with the requirements of DIN 2304
3. Acquisition of optimization potentials and derivation of concrete approaches to solutions.

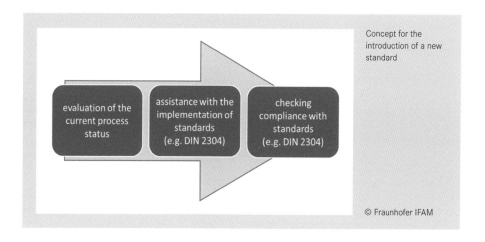

Concept for the
introduction of a new
standard

© Fraunhofer IFAM

Supporting experts

Introducing a new standard to a company's operations is a complex and elaborate process, and many years of experience with adhesive bonding processes and the associated quality assurance are an advantage. If these skills or resources are not available internally, external experts can be brought in.

A specific concept will be developed based on the existing possibilities in the company. To do this as described above, the process status is initially analyzed. In addition to the acquisition of optimization potentials and the derivation of possible solutions, the implementation is actively supported. Manufacturing documents or e.g. the complete process planning is created, and the customer is actively supported while introducing the standard. Finally, the implementation of the measures is analyzed, and a compliance review is carried out to ensure preparation for possible certification. Support in all of these areas can be provided by the Fraunhofer Institute for Manufacturing Technology and Advances Materials IFAM, Division for Adhesive Bonding Technology and Surfaces.

Dipl.-Ing. (FH) Andrea Paul and
Dipl.-Phys. Kai Brune
Fraunhofer Institute for Manufacturing Technology and
Advanced Materials IFAM, Bremen, Germany

www.bremen-bonding.com

COMPANY PROFILES

Adhesive Producer
Raw Material Supplier

3M
Deutschland GmbH

Carl-Schurz-Straße 1
D-41453 Neuss
Phone +49 (0) 21 31-14 33 30
Fax +49 (0) 21 31-14 32 00
E-Mail: kleben.de@mmm.com
www.3M-klebtechnik.de

Member of IVK

Company

Year of formation
1951

Size of workforce in Germany
6,700

3M Deutschland GmbH
3M Headquarter and Customer Technical
Center, Neuss
Manufacturing sites, Hilden, Kamen
European Distribution Center, Jüchen
3M Oral Care Solutions Division, Seefeld,
Landsberg/Lech
3M Separation and Purification Sciences Division,
Wuppertal, Obernburg
3M Health Information Systems (HIS), Berlin

Subsidiaries 3M Deutschland GmbH
3M Services GmbH, Neuss
(with QNG Subsidiary, Hannover)
Dyneon GmbH, Burgkirchen
TOP-Service for Lingualtechnik GmbH, Bad Essen
Wendt GmbH, Meerbusch, Niederstetten, Jena,
Hameln
Winterthur Technology GmbH, Reutlingen
(Subsidiary of Wendt GmbH)

Further 3M Companies in Germany
3M Medica, Neuss
3M Technical Ceramics, Kempten

Sales channels
Distributor + direct

Management
Dr. John Banovetz, Prof. Dr. Joerg Dederichs,
Michael Peters

Range of Products

1-part structural adhesives
2-part structural adhesives

Types of adhesives
Hot melt adhesives
Reactive adhesives
Dispersion adhesives
Solvent-based adhesives
Spray adhesives
Pressure-sensitive adhesives

Types of sealants
Acrylic sealants
Butyl sealants
PUR sealants
SMP/STP sealants

Raw materials
Fillers
Polymers

Equipment, plant and components
for conveying, mixing, metering and for adhesive
application
for surface pretreatment
for adhesive curing
adhesive curing and drying
measuring and testing

For applications in the field of
Paper/packaging
Bookbinding/graphic design
Wood/furniture industry
Construction industry, including floors, walls and
ceilings
Electronics
Mechanical enineering and equipment construction
Automotive industry, aviation industry
Textile industry
Adhesive tapes, labels
Hygiene
Household, recreation and office

Adtracon GmbH
Hofstraße 64
D-40723 Hilden
Germany
Phone +49 (0) 21 03-253 1710
Fax +49 (0) 21 03-253 1719
Email: info@adtracon.de
www.adtracon.de

Member of IVK

Company

Year of formation
2002

Size of workforce
15

Ownership structure
Dr. Roland Heider, ISB, KfW

Sales channels
Direct and distributors

Contact partners
Management:
Dr. Roland Heider

Further information
Adtracon specializes in the development,
production and marketing of reactive hot
melt adhesives. We also offer technical
and laboratory services and consultancy
services.

Range of Products

Types of adhesives
Reactive hot melt adhesives

For applications in the field of
Bookbinding/graphic design
Woodworking/furniture industry
Automotive industry, aviation industry
Textile/filter/shoe and leather industry

Alberdingk Boley GmbH
Düsseldorfer Straße 53
47829 Krefeld, Germany
Phone +49 (0) 21 51-5 28-0
Fax +49 (0) 21 51-57 36 43
Email: info@alberdingk-boley.de
www.alberdingk-boley.de

Member of IVK

Company

Year of formation
1828

Size of workforce
400 (worldwide)

Ownership structure
family-owned

Subsidiaries
Alberdingk Boley, Inc., Greensboro, USA,
Alberdingk Resins (Shenzhen) Co. Ltd.,
Shenzhen, China,
Thai Castor Oil Industries Co., Ltd.,
Bangkok, Thailand

Sales channels
worldwide

Contact partners
Management:
Timm Wiegmann
Sales and Marketing

Application technology and sales:
Irene Tournee,
Technical Marketing Adhesives

Marketing Coatings:
Johannes Leibl, Manager Sales Dispersions

Range of Products

Raw materials
Polymers:
Polyurethane dispersions
Acrylate dispersions
Styrene acrylic dispersions
Vinylacetate copolymer dispersions
UV-curable dispersions

For applications in the field of
Adhesives:
Tapes, Labels,
Paper/Packaging
Construction
Automotive
Foils/flexible packaging

Coatings:
Wood/Furniture
Metal/Plastics
Construction including floors, walls and
ceilings
Textile/Leather
Film coatings
Primer

ALFA Klebstoffe AG

Vor Eiche 10
CH-8197 Rafz
Phone +41 43 433 30 30
Fax +41 43 433 30 33
Email: info@alfa.swiss
www.alfa.swiss

Member of FKS

Company

Year of formation
1972

Size of workforce
53

Ownership structure
Family Private Limited (AG) Company

Subsidiaries
ALFA Adhesives, Inc. (Partner)

Sales channels
International distribution network

Contact partners
Management:
info@alfa.swiss

Application technology and sales:
info@alfa.swiss

Further information
ALFA Klebstoffe AG, an innovative family business, focused on the development, production and distribution of water-borne adhesives and hotmelts; therefore, ALFA Klebstoffe AG offers a significant benefit regarding the economic and ecologic design of the gluing process of their customers.

Range of Products

Types of adhesives
Dispersion adhesives
Hot melt adhesives
Pressure-sensitive adhesives

For applications in the field of
Foam converting industry
Paper/packaging
Bookbinding/graphic design
Wood/furniture industry
Automotive industry, aviation industry
Hygiene

APM Technica AG

Max-Schmidheinystrasse 201
CH-9435 Heerbrugg
Phone +41 (0) 71 788 31 00
Fax +41 (0) 71 788 31 10
Email: info@apm.technica.com
www.apm-technica.com

Member of FKS

Company

Year of formation
2002

Size of workforce
140

Ownership structure
Private Shareholder

Subsidiaries
APM Technica Philippines, APM Technica
GmbH, Abatech, Polyscience AG

Sales channels
Direct & Distributor

Contact partners
Management:
Andreas Hedinger, Bruno Köppel

Application technology and sales:
APM Technica AG is the full-service provider
in the field of adhesives and surface technology.

Further information
www.apm-technica.com

Range of Products

Types of adhesives
Reactive adhesives

Types of sealants
Polysulfide sealants
Silicone sealants
MS/SMP sealants

Equipment, Plant and Components
for surface pretreatment
for adhesive curing
adhesive curing and drying

For applications in the field of
Electronics
Mechanical engineering and equipment
construction
Automotive industry, aviation industry
Medical and optic industry

ARDEX GmbH

Friedrich-Ebert-Straße 45
D-58453 Witten
Phone +49 (0) 2 30 26 64-0
Fax +49 (0) 2 30 26 64-3 75
Email: info@ardex.de
www.ardex.de

Member of IVK, FCIO, VLK

Company

Year of formation
1949

Size of workforce
2,500

Ownership structure
Private

Subsidiaries
Australia, Austria, Bulgaria, China, Czech
Republic, Denmark, Finland, France, Germany,
Hungary, Hong Kong, India, Ireland, Italy,
Korea, Luxembourg, Mexico, New Zealand,
Norway, Poland, Romania, Russia, Singapore,
Spain, Sweden, Switzerland, Taiwan, Turkey,
UK, United Arab Emirates, USA

Contact partners
Management:
Mark Eslamlooy (CEO)
Dr. Ulrich Dahlhoff
Dr. Hubert Motzet

Application technology and sales:
Daniel Händle

Range of Products

Types of adhesives
Dispersion adhesives
Pressure-sensitive adhesives

Types of sealants
Acrylic sealants
Butyl sealants
PUR sealants
Silicone sealants
MS/SMP sealants

For applications in the field of
Construction industry, including floors,
walls and ceilings

ARLANXEO Deutschland GmbH
Chempark Dormagen, Building F41
Alte Heerstraße 2
D-41540 Dormagen
www.arlanxeo.com

Member of IVK

Company

ARLANXEO is a world-leading synthetic rubber company with sales of around EUR 2.7 billion in 2016, about 3,800 employees and a presence at 20 production sites in nine countries.
The company's core business is the development, manufacturing and marketing of synthetic high-performance rubber for use in the automotive and tire industries, the construction industry, and the oil and gas industries.
ARLANXEO was established in April 2016 as a joint venture of LANXESS and Saudi Aramco.

Contact persons
For Baypren® ALX and Levamelt®
Dr. Martin Schneider
High Performance Elastomers
Phone: +49 221 8885 5908
E-Mail: martin.schneider@arlanxeo.com

For X_Butyl® RB
Dr. Thomas Rünzi
Tire & Specialty Rubbers
Phone: +49 221 8885 4829
E-Mail: thomas.ruenzi@arlanxeo.com

Range of Products

Raw materials for the production of adhesives and sealants

Levamelt® and Baypren® ALX
Owing to its special properties both products are excellent suited for various applications in adhesives technologies.

Baypren® ALX:
Polychloroprene for solvent-based contact adhesives.

Levamelt®:
Ethylene-vinyl acetate copolymers with vinyl acetate contents from 40 to 90 % as base polymers for pressure sensitive adhesives and as modifiers for structural adhesives and hot melts.

X_Butyl® RB:
Isobutylene/isoprene copolymers that are highly impermeable, non-staining and have low levels of unsaturation. These copolymers provide excellent heat stability and are suitable for use in the adhesive and sealant industry.

artimelt AG

Wassermatte 1
CH-6210 Sursee
Phone +41 41 926 05 00
Fax +41 41 926 05 29
Email: info@artimelt.com
www.artimelt.com

Member of FKS

Company

Year of formation
2016

Ownership structure
artimelt is owned by the LAS Holding, Collano and nolax are in the same group of companies.

Subsidiaries
artimelt Inc., Tucker, GA 30084, USA

Sales channels
Direct sales and agents

Managing director
Walter Stampfli

Contact partner
Christoph Lang
Phone: +41 41 926 05 28
Email: christoph.lang@artimelt.com

Further information
artimelt develops, produces and sells hot melt adhesives and has worldwide 45 employees. The Centre of Competence is located in Switzerland.

Range of Products

Types of adhesives
Hot melt adhesives

For applications in the field of
Medical products
Labels
Tapes
Packaging
Security systems
Building
Graphic arts

ADHESIVE SYSTEMS

ATP adhesive systems AG
Sihleggstrasse 23
CH-8832 Wollerau
Phone +41 (0) 43 888 15 15
Fax +41 (0) 43 888 15 10
Email: info@atp-ag.com
www.atp-ag.com

Company

Year of formation
1989

Size of workforce
250

Managing partners
Managing Director:
Daniel Heini

Ownership structure
Privately owned

Sales channels
Direct sales channels
Graphic Distributors

Further information
ATP is a leading manufacturer of high quality
technical tape solutions for the automotive,
foam, graphic, label, semi-structural composite, building and construction industries.
With our extensive technical and marketing
knowledge, our passion for developing
customer-focused solutions and our committed employees that go that little bit further,
success with ATP is a given.
ATP is producing high quality single- and
double-sided adhesive tapes on very modern
coating machines in Germany since 1991.
Using a broad range of adhesive and support
materials, ATP produces single- and double-
sided pressure sensitive adhesive tapes,
transfer tapes and heat-seal films. The
adhesive formulations are solvent free and
are exclusively developed by ATP.

Range of Products

Types of adhesives
Dispersion adhesives
Pressure-sensitive adhesives
Self adhesive tapes
(Single- and double-sided tapes with
different carriers)

For applications in the field of
Paper/packaging
Bookbinding/graphic design
Wood/furniture industry, including floors,
walls and ceilings
Electronics
Automotive industry, aviation industry
Textile industry
Adhesive tapes, labels

ATP's production methods meet the most
modern technological requirements. The
products are developed and manufactured
under a quality management system which is
certified according to ISO 9001, ISO 14001,
ISO 50001 and ISO/TS 16949.
ATP always strives to exceed customer ex-
pectations by developing customer aligned
solutions which offer technical advantages,
competitively and quickly.

Avebe U. A.

Prins Hendrikplein 20
NL-9641 GK Veendam
Phone +31 (0) 598 66 91 11
Fax +31 (0) 598 66 43 68
Email: info@avebe.com
www.avebe.com

Member of VLK

Company

Established in
1919

Employees
1,311

Ownership structure
Cooperation of farmers

Sales channels
Direct and through specialized
distributors worldwide

Further information
Avebe U. A. is an international Dutch starch
manufacturer located in the Netherlands and
produces starch products based on potato
starch and potato protein for use in food,
animal feed, paper, construction, textiles and
adhesives.

Range of Products

Raw materials
Dextrins and Starch based adhesives
Starch ethers

For applications in the field of
Paper and packaging
Paper sack adhesive
Tube winding adhesive
Remoistable envelope adhesive
Protective colloid in polyvinyl acetate
based dispersions
Water activated gummed tape adhesive
Wallpaper and bill posting adhesive
Additive – Rheology modifier for cement
and gypsum based mortars and tile
adhesives
Water purification

BASF SE
D-67056 Ludwigshafen
Phone +49 (0) 6 21-60-0
Email: industrial-adhesives@basf.com
pressure-sensitive-adhesives@basf.com
info-pib@basf.com
www.basf.com

Member of IVK, VLK

Company

Year of formation
1865

Size of workforce
approximately 114,000 employees
(as of year end 2016)

Further information
Areas of adhesive technology:
Pressure-sensitive adhesives
Industrial adhesives
Polyisobutene (PIB)

Range of Products for Adhesives
1. Acrylic dispersions (water-based)
2. Acrylic hot melts (UV-curable)
3. Polyurethane Dispersions (PUD)
4. Styrene-butadiene dispersions
5. Polyisobutene (PIB)
6. Polyvinylpyrrolidone (PVP)
7. Polyvinyl (PV)
8. Auxiliaries:
   • Crosslinking agent
   • Defoamer
   • Thickener
   • Wetting agent
9. Additives
   • Antioxidants
   • Light stabilizer
   • Photoinitiators/Curing agents
   • Others

Email contacts
1 – 9: pressure-sensitive-adhesives@basf.
       com; industrial-adhesives@basf.com
5: info-pib@basf.com

Range of Products

Comments
BASF is one of the world's leading manufacturers of raw materials and additives for industrial and pressure-sensitive adhesives.

With its technologies, BASF offers its customers an alternative to traditional fastening methods. The additive portfolio helps to improve products no matter which adhesive technology is used. BASF's polymer dispersions, UV curable hot melts and auxiliaries are an excellent solution for the production of high-quality self-adhesive products such as labels, tapes and films.

For high sophisticated adhesives, BASF assists with know-how, reliability and safety based on decades of experience. BASF constantly develops and tests new products – according to customer requirements.

BASF also supplies a wide range of polyisobutenes (PIB) with low-, medium- and high molecular weight. Beyond other applications, Glissopal® (LM PIB) is used as tackifier to adjust stickiness of adhesive formulations. The applications for Oppanol® (MM and HM PIB) range from adhesives, sealants through to chewing gum base.

BEARDOWADAMS™
Unique Adhesives

Beardow Adams GmbH
Vilbeler Landstraße 20
D-60386 Frankfurt/M.
Phone +49 (0) 69-4 01 04-0
Fax +49 (0) 69-4 01 04-1 15
www.beardowadams.com

Company

Year of formation
1875

Sales channels
direct and agencies

Contact partners
Sales and Marketing Management:
Janet Pohl

Marketing Assistant:
Martina Hartmann

Range of Products

Types of adhesives
Hot melt adhesives
Reactive adhesives
Dispersion adhesives
Casein, dextrin and starch adhesives
Pressure-sensitive adhesives

For applications in the field of
Paper converting/packaging/labelling
Bookbinding/graphic design
Wood/furniture industry
Electronics
Mechanical engineering and equipment
construction
Automotive industry
Adhesive tapes, labels

Berger-Seidle GmbH

Parkettlacke · Klebstoffe · Bauchemie

Maybachstraße 2
D-67269 Grünstadt/Weinstraße
Phone +49 (0) 63 59-80 05-0
Fax +49 (0) 63 59-80 05-50
Email: info@berger-seidle.de
www.berger-seidle.de

Member of IVK

Company

Year of formation
1926

Size of workforce
85

Ownership structure
100 % subsidiary to Phil. Berger GmbH

Sales channels
Distributors and sales partners/representatives on each country

Contact partners
Management & Sales:
Markus M. Adam

Application technology:
Dr. Wolfgang Kahlen

Further information
www.berger-seidle.de

Range of Products

Types of adhesives
Reactive adhesives
Solvent-based adhesives
Dispersion adhesives
PU-adhesives

Types of sealants
Acrylic sealants
PUR sealants
MS/SMP sealants

For applications in the field of
Wood/furniture industry
Construction industry, including floors, walls and ceilings

BLUFIXX GmbH, Lightcuring Systems
Rodenkirchener Straße 200
D-50389 Wesseling
Phone +49 (0) 22 36-33 63 40
Fax +49 (0) 22 36-33 63 411
Email: info@blufixx.com
www.www.blufixx.com

Member of IVK

Company

Year of formation
2012

Size of workforce
8

Managing partner
Dinko Jurcevic

Nominal capital
25.000 €

Ownership structure
BLUFIXX is 100% subsidiary of KDS Holding

Sales channels
Retail, Internet, Wholesale

Contact partners
Management:
CEO Dinko Jurcevic
Email: dinko.jurcevic@blufixx.com

Sales:
Head of Department Sales
Gezim Neziri
Email: gezim.neziri@blufixx.com

Application Technology:
Johannes Kroner
Email: johannes.kroner@blufixx.com

Range of Products

Types of adhesives
Reactive adhesives

Types of sealants
Acrylic sealants

Raw materials
Additives: Photo Initiators
Resins: Acrylic Resin

Equipment, Plant and Components
for conveying, mixing, metering and for
adhesive application
for surface pretreatment
for adhesive curing

For applications in the field of
Wood/furniture industry
Construction industry, including floors, walls
and ceilings
Automotive industry, aviation industry
Household, recreation and office

BODO MÖLLER CHEMIE
Engineer chemistry

Bodo Möller Chemie GmbH
Senefelderstraße 176
D-63069 Offenbach am Main
Phone +49 (0) 69-83 83 26-0
Fax +49 (0) 69-83 83 26-199
Email: info@bm-chemie.de
www.bm-chemie.de

Company

Year of formation
1974

Employees
About 200

Managing partners
Korinna Möller-Boxberger, Frank Haug, Jürgen Rietschle

Ownership structure
Family owned

Regions
Germany, Austria, Slovenia, Switzerland, France, Benelux, Denmark, Sweden, Norway, Finland, Estonia, Poland, Lithuania, Latvia, Czech Republic, Slovakia, Hungary, Croatia, Russia, India, China, Southern Africa, Sub Sahara Region, Egypt, Morocco, Middle East, Israel, USA and Mexico in foundation

Sales channels
Own sales structure on a local base by own sales force

Contact partners
Management:
info@bm-chemie.de

Application technology and sales:
info@bm-chemie.de

Further information
Leading supplier for specialty chemicals and partner for industrial high performance adhesives, casting resins and engineering plastics in Europe, Africa, Asia and America with more than 40 years experience in various applications in all fields of the processing industry. Bodo Möller Chemie has its own adhesives application laboratory and is certified for aviation and railway.

Range of Products

Types of adhesives
Hot melt adhesives, Reactive adhesives, Solvent-based adhesives, Dispersion adhesives, Vegetable adhesives, dextrin and starch adhesives, Pressure-sensitive adhesives, Silicone adhesives, Sprayable adhesives, Methacrylate adhesives, Epoxy adhesives, 1-component adhesives, 2-component adhesives, Anaerobic adhesives, Cyanoacrylates, PUR adhesives

Types of sealants
Acrylic sealants
Butyl sealants
Polysulfide sealants
PUR sealants
Silicone sealants
MS/SMP sealants

Raw materials
Additives: Stabilizers, Antioxidants, Rheological Modifiers, Tackifiers, Thickeners, Dispersing Agents, Flame Retardants, Pigments, HALS, UV Stabilizers, Crosslinkers
Fillers: Aluminium Oxide, Aluminium Hydroxide, Dolomite, Mica, Chalk, Talc, Ground Quartz and Wollastonite, Calcium Carbonate, Kaolin.
Resins: Acrylic Dispersions, Polyurethane Dispersions, Epoxy Resins, Kolophonium Resins, Reactive Diluents, Cobalt-free Drying Agents
Polymers: Fomulated Polymers EP, PU, PA

For applications in the field of
Paper/packaging, Bookbinding/graphic design, Wood/furniture industry, Construction industry, including floors, walls and ceilings, Electronics, Mechanical engineering and equipment construction, Automotive industry, aviation industry, Textile industry, Adhesive tapes, labels, Hygiene, Household, recreation and office

Bona AB
Murmansgatan 130, Box 210 74
S-20021 Malmö
Phone +46 40 38 55 00, Fax +46 40 18 16 43
Email: bona@bona.com

Adhesive Production
Bona GmbH Deutschland
Jahnstraße 12, D-65549 Limburg
Phone +49 (0) 64 31-4 00 80

Member of IVK

Company

Year of formation
1919

Size of workforce
500

Ownership structure
privat owned stock cooperation

Subsidiaries
Austria: Bona Austria GmbH
Phone +43 662 66 19 43-0
Belgium: Bona NV
Phone +32 2 721 2759
Brazil: BonaKemi Pisos de Madeira
do Brasil Ltda
Phone +55 41 3233 5983
Bona AB Branch Panama
Phone +507 227 2799
Bona Trading (Shanghai) Co., Ltd
Phone +86 10 67 72 8301
Czech Republic & Slovak Republic:
Bona CR spol. s.r.o.
Phone CR +420 236 080 211
Phone SR +421 265 457 161
France: Bona France
Phone: +33 3 88 49 18 60
Germany
Bona Vertriebsges. mbH Deutschland
Phone +49 6431 4008 0
The Netherlands: Bona Benelux BV
Phone +31 23 542 1864
Poland: Bona-Polska Sp. z.o.o.
Phone +48 61 816 34 60/61
Romania: Bona S.r.l.
Phone +40 31 405 75 93

Range of Products

Types of adhesives
Reactive adhesives
Dispersion adhesives

For applications in the field of
Construction industry, for wood floors and
LVT's

Singapore: BonaFar East & Pacific Pte Ltd
Phone +65 6377 1158
Spain/Portugal: Bona Ibérica
Phone +34 916 825522
Italy: Biffignandi spa – Via Circonvallazione
Est 2/6 – Cassolnovo (PV) Italia
Phone 0381 920111
Hungary: M.L.S. Magyarország Kft. -
2310 Szigetszentmiklós, Selló u. 8.
HUNGARY - Phone/Fax: (06 24) 525 400
United Kingdom: Bona Limited
Phone +44 1908 525 150
United States: Bona US
Phone +1 303 371 1411

Contact partners
Management
Dr. Thomas Brokamp (Production, R & D)

Product Management:
Torben Schuy, Thomas Hallberg

R&D
Dr. Holger Wickel

Bostik GmbH
An der Bundesstraße 16
D-33829 Borgholzhausen
Phone +49 (0) 54 25-8 01-0
Email: info.germany@bostik.com
www.bostik.de

Member of IVK, VLK

Company

Year of formation
1889

Size of workforce
400

Subsidiaries
MEM Bauchemie GmbH, Bostik Austria

Sales channels
Construction distribution,
industry, DIY

Contact partners
Management:
Olaf Memmen, Managing Director
Richard Riepe, Business Manager Construction
Norbert Uniatowski, Business Manager FlexLam
Frank Mende, R&D Director
Bernd Köhler, Logistic Manager
Gerhard Flottmann, Purchasing Director
Dr. Michael Nitsche, Production Director

Further information
With annual sales of € 1.95 billion, the company employs 6,000 people and has a presence in more than 50 countries. For the latest information, visit www.bostik.com

Range of Products

Types of adhesives
Hot melt adhesives
Reactive adhesives
Solvent-based adhesives
Dispersion adhesives
Vegetable adhesives, dextrin and starch adhesives
Pressure-sensitive adhesives
Polymer modified binders
SMP adhesives

Types of sealants
Acrylic sealants
Butyl sealants
PUR sealants
Silicone sealants
MS/SMP sealants

Raw materials
Additives, Fillers, Resins, Solvents
Polymers, Starch

For applications in the field of
Paper/packaging
Bookbinding/graphic design
Wood/furniture industry
Construction industry, including floors, walls and ceilings
Mechanical engineering and equipment construction
Automotive industry, aviation industry
Textile industry
Hygiene
Household, recreation and office
Flexible Laminating

Botament
Systembaustoffe
GmbH & Co. KG

Tullnerstraße 23
A-3442 Langenrohr
Phone +43 (0) 22 72-6 74 81
Fax +43 (0) 22 72-6 74 81-35
Email: info@botament.at
www.botament.at

Member of IVK, FCIO

Company

Year of formation
1993

Size of workforce
14

Ownership structure
GmbH & Co. KG

Sales channels
wholesale

Contact partners
Management:
Prok. Ing. Peter Kiermayr

Application technology and sales:
Karl Prickl

Range of Products

Types of adhesives
Reactive adhesives
Dispersion adhesives
Tile adhesives/natural stone adhesives

Types of sealants
Silicone sealants

For applications in the field of
Construction industry, including floors walls
and ceilings

Brenntag GmbH
Stinnes-Platz 1
45472 Mülheim / Ruhr
Phone +49 (0) 208/7828-0
Fax +49 (0) 208/7828-7530
Email: alain.kavafyan@brenntag.de
www.brenntag-gmbh.de

Member of IVK

Company

Year of formation
1874

Size of workforce
1,200

Managing partners
Matthias Compes,
Roland Saenger,
Cosimo Alemano

Nominal capital
154.5 Mio. Euro (Brenntag AG)

Ownership structure
Listed on the stock exchange (Brenntag AG)

Contact partners
Management:
Alain Kavafyan

Application technology and sales:
Michael Hesselmann
Markus Wolff

Range of Products

Raw materials
Additives:
Accelerators, Adhesion Promoters, Antioxidants, Biocides, Catalysts, Defoamers, Dispersing Agents, Matting Agents, Plasticizers, Polyether Amines, Rheology Modifiers, Silanes, Surfactants, Thickeners, UV-Stabilisers

Resins:
Acrylic Dispersions and Resins, Acrylic Monomers, Styrene Acrylics
Epoxy Resins incl. Curing Agents, Reactive Diluents and Modifiers,
Hydrocarbon Resins
PU-Systems: Polyols and Isocyanates (aromatic and aliphatic),
Prepolymers, PUR-Dispersions
Silikone Resins and Emulsions

Solvents: all kinds of solvents

Polymers: PMMA, Silicones

For applications in the field of
Paper/packaging
Bookbinding/graphic design
Wood/furniture industry
Construction industry, including floors, walls and ceilings
Electronics
Automotive industry, aviation industry
Textile industry
Adhesive tapes, labels
Composites

)(BÜHNEN

Bühnen GmbH & Co. KG
Hinterm Sielhof 25
D-28277 Bremen
Phone +49 (0) 4 21-51 20-0
Fax +49 (0) 4 21-51 20-2 60
Email: info@buehnen.de
www.buehnen.de

Member of IVK

Company

Year of formation
1922

Size of workforce
89

Ownership structure
Private ownership

Subsidiaries
BÜHNEN Polska Sp. z o. o.
BÜHNEN B. V., NL
BÜHNEN, AT

Contact person
Managing Director:
Bert Gausepohl

International Sales / Marketing
Valentino di Candido

Sales GER, AT, CH
Hans-Gerhard Hartje

Distribution channels
Direct Sales, Distributors

Range of Products

Hot Melt Adhesives
The product range includes a
variety of different hot melt adhesives
for almost every application.
Available bases:
EVA, PO, POR, PA, PSA, PUR, Acrylate.
Available shapes:
slugs, sticks, granules, pillows, blocks,
cartridges, barrels, drums, bags.

Application Technology
Hot melt tank applicator systems with
piston pump or gear pump, PUR- and POR-
hot melt tank systems, PUR- and POR-bulk
unloader, hand guns for spray and line
application, application heads for line, slot,
spray, dot, spiral application and special
application heads for individual customer
requirements, hand-operated glue applicators, PUR- and POR glue applicators, wide
range of application accessories, customer-
oriented application, solutions.

Applications
Automotive, Packaging, Display Manufac-
turing, Electronic Industry, Filter Industry,
Shoe Industry, Foamplastic and Textile
Industry, Case Industry, Construction
Industry, Florists, Wood-Processing and
Furniture Industry, labelling Industry.

BYK Additives & Instruments
Abelstraße 45
D-46483 Wesel
Phone +49 (0) 281-670-0
Fax +49 (0) 281-6 57 35
Email: info@byk.com
www.byk.com

Member of IVK

Company

Year of formation
1962

Size of workforce
Around 2,100 people worldwide

Managing partners
Dr. Stephan Glander (President),
Albert von Hebel (Finance),
Gerd Judith (Managing Director Asia)

Ownership structure
BYK is a member of ALTANA, Germany

Subsidiaries
BYK-Chemie (Germany), BYK (Brazil), BYK Additives (China), BYK (India), BYK Japan (Japan), BYK Korea (Korea), BYK Netherlands (Netherlands), BYK Chemie de México (México), BYK Asia Pacific (Singapore, Taiwan, Thailand and Vietnam), BYK Additives (United Kingdom), BYK (U.A.E.), BYK USA (USA), PolyAd Services (USA)
- Warehouses and representations in > 100 countries and regions
- Technical Service Labs in Germany, Brazil, China, India, Japan, Korea, México, The Netherlands, Singapore, U.A.E., United Kingdom and in the USA
- Production Sites in Germany, China, The Netherlands, United Kingdom and in the USA

Sales channels
Worldwide – direct (BYK) and indirect (agents and distributors)

Close to the customer
BYK places great importance on being close to its customers and remaining in constant dialog with them. This is one of the reasons why the company is represented in more than 100 countries and regions around the globe. In over 30 technical service laboratories, BYK offers customers and application engineers support for concrete questions.

Range of Products

Raw materials
Additives: Wetting and dispersing additives, rheological additives (PU thickener, organically modified clays, synthetic and natural layered silicates), defoamers and air release agents, additives to improve substrate wetting, surface slip and levelling, UV-absorbers, wax additives, anti-blocking additives, conductive additives, nano based additives, adhesion promoters

For applications in the field of
Paper/packaging
Bookbinding/graphic design
Wood/furniture industry
Construction industry including floors, walls and ceilings
Electronics
Sealants
Automotive industry, aviation industry
Textile industry
Adhesive tapes, labels
Hygiene
Household, recreation and office

Contact partners for your customers
Mr. Tobias Austermann,
Email: Tobias.Austermann@altana.com
Phone: +49 281-670-28128

Further information about your company
BYK Additives & Instruments is one of the world's leading suppliers in the field of additives and measuring instruments.The coatings, inks and plastics industries are among the main consumers of BYK additives. Yet with oil production, the manufacture of care products, paper surface finishing, the manufacture of adhesives and sealants, or construction chemistry, too, BYK additives improve the product characteristics and production processes.

Byla GmbH

Industriestraße 12
D-65594 Runkel
Phone +49 (0) 64 82-91 20-0
Fax +49 (0) 64 82-91 20-11
Email: contact@byla.de
www.byla.de

Member of IVK

Company

Year of formation
1975

Nominal capital
90,000 €

Sales channels
Worldwide

Range of Products

Types of adhesives
Reactive adhesives

For applications in the field of
Wood/furniture industry
Electronics
Mechanical engineering and equipment
construction
Automotive industry, aviation industry

Celanese Europe BV

The New Atrium, 6th floor
Strawinskylaan 3105
1077 ZX Amsterdam
Phone +31 (0) 20721 3700
Email: mowilith.info@celanese.com
www.celanese.com/emulsion-polymers

Member of IVK

Company

Year of formation
2003 Emulsions Business

Size of workforce
7,200 (Celanese Group)

Ownership structure
Celanese Corporation

Contact partners
Management:
Robert van Zundert

Application technology and sales:
Dr. Bernhard Momper
Stefan Wolfsheimer
Christian Lindemann

Range of Products

Types of adhesives
Dispersion for adhesives
water based Dispersions

Raw materials
Emulsion Polymers

For applications in the field of
Paper/packaging
Bookbinding/graphic design
Wood/furniture industry
Construction industry, including floors,
walls and ceilings
Automotive industry, aviation industry
Textile industry

certoplast
Technische Klebebänder
GmbH

Müngstener Straße 10
D-42285 Wuppertal
Phone +49 (0) 20 2-2 55 48-0
Fax +49 (0) 20 2-2 55 48-48
Email: verkauf@certoplast.com
www.certoplast.com

Member of IVK
Member of Afera

Company

Year of formation
1991

Size of workforce
50

Managing partners
P. Rambusch
Dr. R. Rambusch

Subsidiaries
certoplast (Suzhou) Co., Ltd., China
certoplast North America Inc., Las Cruses

Contact partners
Management:
P. Rambusch (General manager)
Dr. R. Rambusch (General manager)

Application technology and sales:
Dr. Andreas Hohmann (Sales manager)

Range of Products

Types of adhesives
Hot melt adhesives
Dispersion adhesives
Pressure-sensitive adhesives

Equipment, plant and components
for adhesive curing
adhesive curing and drying

For applications in the field of
Construction industry, including floors,
walls and ceilings
Electronics
Automotive industry, aviation industry
Adhesive tapes, labels

Chemetall
expect more +

now part of BASF Group

Chemetall GmbH
Trakehner Straße 3
D-60487 Frankfurt/M.
Phone +49 (0) 69-71 65-0
Fax +49 (0) 69-71 65-29 36
www.chemetall.com

Member of IVK

Company

Year of formation
1982

Managing partners
Board of Management:
Dr. M. Jung

Ownership structure
A Global Business Unit of BASF's
Coatings division

Subsidiaries
> 40 worldwide

Sales channels
CM subsidiaries and specific distributors

Contact partners
Management:
Thomas Willems

Application technology and sales:
Ralph Hecktor
Phone +49 (0) 69-71 65-24 46

Further information
See website: www.chemetall.com
Certification to ISO 9001, EN 9100,
ISO 14001

Range of Products

Types of adhesives
Hot melt adhesives
Reactive adhesives

Types of sealants
Polysulfide sealants
PUR sealants
Other epoxy

For applications in the field of
Electronics
Mechanical engineering and equipment
construction
Automotive industry, Aviation industry

Chemische Fabrik Budenheim KG

Rheinstraße 27
D-55257 Budenheim
Phone +49 (0) 6139-89-0
Email: info@budenheim.com
www.budenheim.com

Member of IVK

Company

Year of formation
1908

Size of workforce
1.000

Managing partners
Dr. Harald Schaub
Dr. Stefan Lihl

Ownership structure
Belongs to the Oetker Group, privately owned

Contact partners
Management:
Email: coatings@budenheim.com

Applications technology and sales:
Email: coatings@budenheim.com

Furher information
www.budenheim.com/clip4coatings

Range of Products

Raw materials
Additives
Fillers
Flame Retardants

For applications in the field of
Wood/furniture industry
Construction industry, including floors, walls and ceilings
Automotive industry, aviation industry
Textile industry

CHT R. Beitlich GmbH

Bismarckstraße 102
72072 Tübingen
Germany
Phone +49 7071 154-0
Fax +49 7071 154-290
Email: info@cht.com
www.cht.com

Member of IVK

Company

Year of formation
1953

Size of workforce
1,900 worldwide

Sales channels
More than 20 CHT affiliates and agencies
worldwide

Management
Dr. Frank Naumann (CEO)
Dr. Bernhard Hettich (COO)
Jan Siebert (CFO)

Contact partners
Application technology and sales:
Eric Knehr (General Industries)
Dennis Seitzer (Textile, F&E Polymers)

Range of Products

Types of adhesives
Powder adhesives
Reactive adhesives
Solvent-based adhesives
Dispersion adhesives
High solid adhesives
Pressure-sensitive adhesives
Acrylic adhesives
PUR adhesives
Silicone adhesives
Thermo-activated adhesives

Types of sealants
Silicone sealants

Raw materials
RTV-1/RTV-2
LSR silicones
Acrylic dispersions
PU dispersions

Additives:
Adhesion promoters and primers
Rheology additives and thickeners
Release agents
Crosslinker, chain extender

For applications in the field of
Construction industry
Electronics
Mechanical engineering and equipment
Automotive industry, aviation industry
Textile industry/Technical textiles
Paper and packaging
Pressure sensitive tapes and labels
Flock
Mould making

CnP Polymer GmbH

Schulteßdamm 58
D-22391 Hamburg
Phone +49 (0) 40-53 69 55 01
Fax +49 (0) 40-53 69 55 03
Email: cnp.polymer@t-online.de
www.cnppolymer.de

Member of IVK

Company

Year of formation
1999

Ownership structure
private

Sales channels
own sales force

Contact
Christoph Niemeyer

Range of Products

Raw materials
Polymers:
SIS, SBS, SEBS
PIB
Acrylic hotmelt polymer „Acrynax"
PTFE, PFA, MFA, ETFE, FEP
IR/SBR/SSBR

Resins:
Hydrocarbon resins C5/C9
Gumrosin derivatives

Adhesives:
„Micronax" Repositionabel

For applications in the field of
Paper/packaging
Bookbinding/graphic design
Wood/furniture industry
Construction industry, including floors,
walls and ceilings
Automotive industry, aviation industry
Textile industry
Adhesive tapes, Labels
Hygiene
Household, recreation and office

Coim Deutschland GmbH
Novacote Flexpack Division
Schnackenburgallee 62
D-22525 Hamburg
Phone +49 (0) 40-85 31 03-0
Fax +49 (0) 40-85 31 03-69
Email: info@coimgroup.com
www.coimgroup.com

Member of IVK

Company

Year of formation
as COIM Group in 1962

Ownership structure
private owned company

Subsidiaries
COIM operates through a network of production sites, commercial companies and
agencies located all over the world

Sales channels
The Novacote Division is part of the COIM
Group, dedicated to developing and supply-
ing adhesives and coatings, mainly for the
Flexible Packaging market.

Contact partners
Management:
Frank Rheinisch

Application technology:
Oswald Watterott

Sales:
Joerg Kiewitt

Further information
The Novacote Division is part of the Coim
Group, dedicated to developing and supp-
lying adhesives and coatings, mainly for the
Flexible Packaging market. During recent
years the Novacote Division grew rapidly
both in terms of Business and Organization.
With respect to the global organization the
Novacote Technology Center is located
in Hamburg, Germany as R&D Centre for
Packaging.

Range of Products

Types of adhesives
Reactive adhesives
Solvent-based adhesives
Dispersion adhesives

Types of sealants
Acrylic sealants
Other

Raw materials
Resins
Polymers

For applications in the field of
Paper/packaging
Adhesive tapes, labels
Hygiene
Insulation materials
Solar panels

Collall B.V.

Electronicaweg 6
NL – 9503 EX Stadskanaal
Phone +31 (0) 599-65 21 90
Fax +31 (0) 599-65 21 91

Member of VLK

Company

Year of formation
1949

Size of workforce
25

Ownership structure
family-owned

Contact partners
Management:
Patrick van Rhijn

Range of Products

Types of adhesives
Solvent-based adhesives
Dispersion Adhesives
Vegetable adhesives, dextrin and
starch adhesives

For applications in the field of
Household, hobby and office
Bookbinding/graphis design
Wood/furniture industry

Additional
supplier of various creative materials
for school and hobby

Collano AG
CH-6203 Sempach Station
Phone +41 41 469 92 75
Fax +41 41 469 93 68
Email: info@collano.com
www.collano.com

Member of FKS

Company

Year of formation
1947

Size of workforce
19

Sales channels
Direct sales, agents or distributors

Contact partners
Management:
Mike Gabriel
Hans Stalder

Sales:
Mike Gabriel
Phone: +41 41 469 92 75

Range of Products

Types of adhesives
Hot melt adhesives
Reactive adhesives
Dispersion adhesives

For applications in the field of
Wood/furniture industry
Construction industry

Coroplast Fritz Müller GmbH & Co. KG

Wittener Straße 271
D-42279 Wuppertal
Phone +49 (0) 2 02-26 81-0
Fax +49 (0) 2 02-26 81-3 80
Email: coroplast@coroplast.de
www.coroplast.de

Member of IVK

Company

Year of formation
1928

Size of workforce
6,200

Sales channels
Wholesale and industry

Contact partners
Management:
N. Mekelburger
M. Söhngen
W. Berns
T. Kämmerer

Further information

Coroplast acts in 3 business units:
• tapes
• cables & wires
• cable assemblies

Range of Products

For applications in the field of
Paper/packaging
Wood/furniture industry
Construction industry, including floors, walls and ceilings
Electronics
Mechanical engineering and equipment construction
Automotive industry, aviation industry
Household, recreation and office

Covestro AG
D-51373 Leverkusen
Phone +49 (0) 214-6009 7184
Email: adhesives@covestro.com
www.adhesives.covestro.com

Member of IVK

Company

Year of formation
2015

Size of workforce
15,600

Contact partners
Marketing Europe/Business Development
Phone +49 (0) 214-6009 7184
Email: adhesives@covestro.com
www.adhesives.covestro.com

Range of Products

Raw materials
Polyurethane-Dispersions (Dispercoll® U)
Hydroxylpolyurethanes (Desmocoll®, Desmomelt®)
Polyisocyanates (Desmodur®)
Isocyanate-Prepolymers (Desmodur®, Desmoseal®)
Silanterminated Polyurethanes (Desmoseal® S)
Polyesterpolyols (Baycoll®)
Polyetherpolyols (Desmophen®, Acclaim®)
Polychloroprene-Dispersions (Dispercoll® C)
Halogenated Polyisoprenes (Pergut®)
Silicon dioxide-nanoparticle dispersions (Dispercoll® S)

INVENTING
POLYMER STANDARDS
FOR YOU

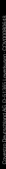

Covestro is an independent, globally leading provider of polymer solutions. In the field of coatings, adhesives, and specialties, we operate an international network of cutting-edge production plants. By producing the same high standard at every site throughout the world, we ensure that our customers are able to apply their formulations globally standardized. For efficiency and reliable quality in every region.

What can we invent for you? **www.inventing-for-you.com**

www.covestro.com

cph Deutschland Chemie GmbH

Heinz-Bäcker-Straße 33
D-45356 Essen
Phone +49 (0) 2 01- 81 40 60
Email: service@cph-group.com
www.cph-group.com

Member of IVK

Company

Year of formation
1975

Size of workforce
ca. 300 (cph group)

Managing partners
CEO Dr. Gerwin Schüttpelz

Subsidiaries
Italy, Turkey, Russia, Ukraina, Portugal

Range of Products

Types of adhesives
Hot melt adhesives
Dispersion adhesives
Vegetable adhesives, dextrin and starch adhesives
Pressure-sensitive adhesives

For applications in the field of
Labelling
Paper/packaging
Wood/furniture industry
Adhesive tapes, labels
Hygiene

CTA GmbH
Industriestraße 2
D-74321 Bietigheim-Bissingen
Phone +49 (0) 71 42-506-0
Fax +49 (0) 71 42-506-1 40

Standort Ludwigsburg:
Voithstraße 1
D-71640 Ludwigsburg
Phone +49 (0) 71 41-29 99 16-0

Email: info@cta-gmbh.de
www.cta-gmbh.de

Member of IVK

Company

Year of formation
2005

Size of workforce
170

Managing partners
Martin Kummer

Ownership structure
Member of Tubex Holding GmbH

Sales channels
direct

Contact partners
Management:
Martin Kummer

Application technology and sales:
Sales: Franco Menchetti
R + D: Dr. Dirk Buchholz

Further information
The core business of CTA GmbH is a wide
range of contract services in the fields of:
- Product manufacturing which includes the
  developing or improvement of composites
  and the mixing of products according to
  customers' recipes and parameters
- The filling of almost all viscous fluid chemical
  products into various kinds of primary and
  secondary packaging – like all kind of tubes,
  cartridges, bottles, pouches, cartons, blister
  packs and others – which are released
  and approved to the best suitability of the
  product and its adequate packaging.

Range of Products

Types of adhesives
Solvent-based adhesives
Dispersion adhesives
Glutine glue
Pressure-sensitive adhesives

Raw materials
Additives, Fillers, Resins, Solvents

Types of sealants
Silicone sealants
MS/SMP sealants
Other

Equipment, plant and components
measuring and testing

For applications in the field of
Paper/packaging
Wood/furniture industry
Construction industry, including floors, walls
and ceilings
Electronics
wind energy
Mechanical engineering and equipment
construction
Automotive industry, aviation industry
Household, recreation and office

- Developing the appropriate packaging in
  accordance to the needs of the product and
  its marketing aspects.
- The packaging assembling to the point of
  sale, supply and distribution logistics offers
  a full contract service package to different
  branches of industries and businesses.

Cyberbond Europe GmbH
A H.B. Fuller Company
Werner-von-Siemens-Straße 2
D-31515 Wunstorf
Phone +49 (0) 50 31-95 66-0
Fax +49 (0) 50 31-95 66-26
Email: info@cyberbond.de
www.cyberbond.eu

Member of IVK

Company

Year of formation
1999

Size of workforce
18

Managing partners
Ulrich Lipper, Holger Bleich, James East,
Bob Martschin

Nominal capital
50,000 EUR

Ownership structure
H.B. Fuller

Subsidiaries
Cyberbond France SARL, France
Cyberbond Iberia, Spain
Cyberbond CS s. r. o., Czech Republic

Sales channels
Direct to the industry and via exclusive
nationwide distributors as well as special
Private Label accounts

Contact partners
Management:
Ulrich Lipper, Holger Bleich

Application technology and sales:
Dr. Lars Hoyer, Ulrich Lipper

Further information
Cyberbond –
The Power of Adhesive Information
IATF 16949
ISO 13485
ISO 9001
ISO 14001

Range of Products

Offered adhesives
Cyanoacrylates
Anaerobic Adhesives and Sealants
UV and Light Curing Adhesives
Additional programme consisting of:
Primers, Activators, D-Bonders and
Dosing Aids

Dosing equipment
LINOP Modular Dosing System for
1K Reactive Adhesives
LINOP UV LED Curing System

Products are used in
Automotive and automotive sub supplier
industry
Electronic industry
Aviation industry
Elastomer/plastic/metal working industry
Machine tool industry
Medical industry
Shoe industry
DIY, Hobby and office

DEKA
Kleben & Dichten
GmbH (Dekalin®)

Gartenstraße 4
D-63691 Ranstadt
Phone +49 (0) 60 41-82 03 80
Fax +49 (0) 60 41-82 12 22
Email: info@dekalin.de
www.dekalin.de

Member of IVK

Company

Year of formation
1907 DEKALIN
1999 DEKA

Size of workforce
together > 125

Ownership structure
family owned

Sales channels
industry, manufacturing,
technical wholesalers

Contact partners
Management:
Michael Windecker

Range of Products

Types of adhesives
Solvent-based adhesives
Dispersion adhesives

Types of sealants
Butyl sealants
PUR sealants
MS/SMP sealants

For applications in the field of
Wood/furniture industry
Construction industry, including floors,
walls and ceilings
Mechanical engineering and equipment
construction
Automotive industry, aviation industry
Household, recreation and office
Caravan, camper and mobile home,
HVAC, Air duct sealing

DELO Industrial Adhesives
DELO-Allee 1
D-86949 Windach
Phone +49 (0) 81 93-99 00-0
Fax +49 (0) 81 93-99 00-1 44
Email: info@DELO.de
www.DELO-adhesives.com

Member of IVK

Company

Year of formation
1961

Size of workforce
560

Managing board
Dr. Wolf-Dietrich Herold
Sabine Herold
Robert Saller

Subsidiaries
Global distributors, subsidiaries in the USA, China and Singapore and representative offices in Japan, Taiwan, Malaysia and South Korea.

Sales channels
Traders and direct

Contact partners
Application technology and sales:
Robert Saller, Managing director

Further information
DELO is a leading manufacturer of industrial adhesives with its headquarters in Windach near Munich, Germany, and subsidiaries in the USA, China and Singapore. In the fiscal year 2017, 560 employees generated sales revenues of EUR 95 million, for 2018, Eur 126 million are expected. The company supplies customized special adhesives and associated technology for high-tech industries such as automotive, aviation, optoelectronics, and electronics. DELO's customers include Bosch, Daimler, Infineon, Osram, Siemens and Sony.

Range of Products

Types of adhesives
Dual-curing adhesives
Light-curing and light-activated acrylates and epoxies
One- and two-component epoxy resins
Electrically conductive adhesives
Methacrylates
Cyanoacrylates
Polyurethanes

Types of sealants
Acrylic sealants
PUR sealants
Silicone sealants
Other

Equipment, plant and components
Spot and area curing lamps
Jet-valves for micro-dispensing
Flexible foil cartridges for bubble-free dispensing

For applications in the field of
Automotive
Consumer and industrial electronics
Optics and optoelectronics
Mechanical engineering
Aviation
White goods

 **DKSH**

DKSH GmbH
Baumwall 3
20459 Hamburg
Phone +49 (0) 40-37 47-3 40
Fax +49 (0) 40-37 47-3 49 33
Email: info.ham@dksh.com
www.dksh.de

Member of IVK

Company

Year of formation
1992

Size of workforce
35

Managing Director
Thomas Sul

Ownership structure
DKSH Group

Subsidiaries of DKSH Group
150/30,320 employees/
net sales CHF 10.5 billion

Sales channels
Own sales force

Sales Manager
Sven Thomas

Further information
DKSH is the leading Market Expansion Services provider with a focus on Asia. As the term „Market Expansion Services" suggests, DKSH helps other companies and brands to grow their business in new or existing markets. Publicly listed on the SIX Swiss Exchange since March 2012, DKSH is a global company headquartered in Zurich. With 150 distribution centers in 35 countries and 30.320 specialized staff, DKSH generated net sales of CHF 10.5 billion in 2016. The company offers a tailor-made, integrated portfolio of sourcing, marketing, sales, distribution, and after-sales services. It provides business partners with expertise as well as on-the-ground logistics based on a comprehensive network of unique size and depth.

Range of Products

Raw materials
Wide range of specialties for Epoxi and PU
Aliphatic isocyanates
High-molecular weight co-polyesters
Liquid isoprene rubbers
Moisture scavengers (PTSI)
Oxetanic reactive diluents and plasticizers
Resins: Co- Polyester, Vinyl, Polyamide-Imide, Acrylic, Ketone, Oxetanic Resins
Heat seal lacquers
Adhesion promoters and primers Polyolefin, Polyolefindispersions, Polyester Lacquers
Plasticizers
Electrical and thermal conductive additives

For applications in the field of
Paper/Pharma/Food packaging Bookbinding/ graphic design Wood/furniture industry
Construction industry, including floors, walls and ceilings
Electronics
Automotive industry, aviation industry
Adhesive tapes, labels
Hygiene

Business activities are organized into four specialized Business Units that mirror DKSH fields of expertise: Consumer Goods, Healthcare, Performance Materials, and Technology. With strong Swiss heritage, the company has now a over150-year-long tradition of doing business in and with Asia, and is deeply rooted in communities and businesses across Asia Pacific.

Dow Deutschland Anlagengesellschaft mbH

Dow Deutschland Inc.

Am Kronberger Hang 4
65824 Schwalbach/Ts
Telefon: +49 (0) 61 96-566-0
www.dow.com

Member of IVK

Company

Foundation
1897 (Dow Chemical)

Size
54,000 employees worldwide (Dow Chemical)

Ownership structure
The Dow Chemical Company

Contact
Dow Customer Information Center
Phone + 800-3-694-6367 Toll Free
Phone + 31-115-67-26-26
Fax + 31-115-67-28-28

General: About Dow Chemical
Dow combines the power of science and technology to passionately innovate what is essential to human progress. The Company connects chemistry and innovation with the principles of sustainability to help address many of the world's most challenging problems such as the need for clean water, renewable energy generation and conservation, and increasing agricultural productivity. Dow's diversified industry-leading portfolio of specialty chemicals, advanced materials, agrosciences and plastics businesses delivers a broad range of technology-based products and solutions to customers in approximately 160 countries and in high growth sectors such as electronics, water, energy, coatings and agriculture. In 2012, Dow had annual sales of $ 56.8 billion and employed approximately 54,000 people worldwide. The Company's more than 5,000 products are manufactured at 188 sites in 36 countries across the globe. The Company conducts its worldwide operations through global businesses, which are reported in six operating segments: Electronic and Functional Materials, Coatings and Infrastructure Solutions, Agricultural.

Range of Products

Dow Monomers
Acrylic Monomers or Acrylic Acid and Esters are key materials used in dispersions and emulsions for adhesives, coatings, inks, woven and non-woven textiles, plastics and polymers and superabsorbent products.

Functional Monomers (GMA, HEA, HEMA and HPA, HPMA) are essential components of automotive coatings, powder coatings, radiation curable coatings, waterborne coatings, industrial and protective finishes, appliance and hardware finishes, adhesives and electrical laminates.

Specialty Monomers - Divinylbenzene (DVB) and Vinylbenzyl Chloride (VBC) – have been used in the synthesis and manufacture of plastics, composites, ion exchange resins, and latexes for coatings and resins in a multitude of applications.

Vinyl Acetate Monomer (VAM for emulsion polymers, resins, and an intermediate used in paints, adhesives, coatings, textiles, wire and cable compounds, laminated safety glass, packaging, automotive plastic fuel tanks, and acrylic fibers.

Dow Microbial Control
Products, application knowledge and local support to control microbial contamination in adhesives based on starch, protein, natural and synthetic gums, rubbers and latexes.

Key segments:
In can Preservation
Dry film Protection
Industrial Hygiene

Dow Automotive Systems
Adhesives and bonding agents
Epoxy resin adhesives
Polyurethane adhesives
Acrylate adhesives
MS Polymers
Primer systems
Bonding agents and flocking systems
Application areas: Vehicle construction, body construction, glass bonding and body repair and glass replacement

Drei Bond GmbH
Carl-Zeiss-Ring 17
85737 Ismaning, Germany
Phone +49 (0) 89-962427 0
Fax +49 (0) 89-962427 19
Email: info@dreibond.de
www.dreibond.de

Member of IVK

Company

Year of formation
1979

Number of employees
48

Partners
Drei Bond Holding GmbH

Share capital
€ 50,618

Subsidiaries
Drei Bond Polska sp. z o.o. in Kraków

Distribution channels
Directly to the automotive industry
(OEM + tier 1/tier 2); indirectly
via trading partners as well as select private
label business

Contacts
Management:
Mr. Thomas Brandl

Application engineering, adhesive and
sealants:
Christian Eicke

Application engineering, metering technology:
Sebastian Schmid, Norbert Frank,
Marco Hein

Adhesive and sealant sales:
Christian Eicke

Metering technology sales:
Norbert Frank, Marco Hein

Additional information
Drei Bond is certified according to ISO 9001-
2015 and ISO 14001-2015

Range of Products

Types of adhesives/sealants
• Cyanoacrylate adhesives
• Anaerobic adhesives and sealants
• UV-light curing adhesives
• 1C/2C epoxy adhesives
• 2C MMA adhesives
• 1C/2C PUR adhesives
• 1C MS hybrid adhesives and sealants
• 1C synthetic adhesives and sealants
• 1C silicone sealants

Complementary products:
• Activators, primers, cleaners

Equipment, systems and components
• Drei Bond Compact metering systems →
  semi-automatic application of adhesives and
  sealants, greases and oils
  Metering technology: pressure/time and
  volumetric
• Drei Bond Inline metering systems →fully
  automated application of adhesives and
  sealants, greases and oils
  Metering technology: pressure/time and
  volumetric
• Drei Bond metering components:
  Container systems: tanks, cartridges, drum
  pumps
  Metering valves: progressive cavity pumps,
  diaphragm valves, pinch valves, spray
  valves, rotor spray

For applications in the following fields
• Automotive industry/automotive suppliers
• Electronics industry
• Elastomer/plastics/metal processing
• Mechanical and apparatus engineering
• Engine and gear manufacturing
• Enclosure manufacturing (metal and plastic)

Dymax Europe GmbH
Kasteler Straße 45
D-65203 Wiesbaden
Phone +49 (0) 611-962 7900
Fax +49 (0) 611-962 9440
Email: info_DE@dymax.com
www.dymax.com

Member of IVK

Company

Year of formation
1995

Size of workforce
250 + worldwide

Ownership structure
Dymax Corporation, USA

Sales channels
Direct and Distributors

Contact partners
Managing Director:
Christoph Gehse

Technical Manager:
Wolfgang Lorscheider

Further information
At Dymax we combine our product offering
of oligomers, adhesives, coatings, dispensing
systems, and curing equipment with our
expert knowledge of light-cure technology.

Range of Products

Types of adhesives
UV and light-curable adhesives
Temporary masking resins
Conformal coatings
Potting materials
Encapsulants
FIP/CIP- gaskets
In addition: Materials with secondary
moisture- and heat cure options, adhesive
activators.

For applications in the field of
Medical
Orthopedic implants
Electronics
Automotive
Aerospace
Optics
Glass
Optical Bonding

Additional Products
UV-Spot and flood lamps (Broadband and LED)
UV-Conveyors
Radiometers
Dispensing equipment
Technical consulting

KLEBSTOFFE • ADHESIVES

Eluid Adhesive GmbH
Heinrich-Hertz-Straße 10
D-27283 Verden
Phone +49 (0) 42 31-3 03 40-0
Fax +49 (0) 42 31-3 03 40-17
Email: info@eluid.de
www.eluid.de

Member of IVK

Company

Year of formation
1932

Size of workforce
8

Managing partners
Andreas May

Ownership structure
100 % Private

Contact partners
Andreas May
Karin Münker

Sales channels
Europe: through our own sales force,
traders and agents
Worldwide: traders and agencies

Range of Products

Types of adhesives
Dispersion adhesives
Dispersion/Pressure-sensitive adhesives
Dextrin, Casein and Starch adhesives
Latex adhesives
PUD Adhesives
PUR Hotmelts
APAO, EVA-, PSA, PO- Hotmelts

For applications in the field of
Bookbinding/graphic-arts
Coatings for films (Soft Touch)
Paper and Converting Envelope industry
Overlaminating films
Heat Seal Products
Packagingindustry
Pressure-sensitive adhesives for film,
Protection of books
Tapes and Labels
Film laminating
Protective film adhesives
Safty Documents
Nowowen industry
Wallpaper industry
Textile industry
Transformerboards

Emerell AG
Eichenstrasse 12
6203 Sempach Station
Switzerland
Phone +41 41 469 91 00
Fax +41 41 469 91 12
Email: info@emerell.com
www.emerell.com

Member of FKS

Company

Emerell is the first independent manufacturing partner for industrial manufacturers, processors and distributors of polymer specialties and high-quality adhesives. As a full custom manufacturer, the company offers a wide range of services to its customers and accompanies them from the first test onto the market launch and further development of their products.

Managing Director
Adrian Leumann

Contact
Norbert Bazelli
Phone +41 41 469 93 13
Email: norbert.bazelli@emerell.com

Subsidiary companies
Emerell Extrusion AG, Schmitten, Switzerland
Emerell GmbH, Buxtehude, Germany

Range of Products

Technologies
Blown film extrusion
Cast film extrusion
Extrusion coating
Polymerisation
Chemical reactions
Mixing techniques for liquid, paste-like and
Reactive products

Services
Process engineering
Manufacture and processing
Filling, packing, labelling
Purchase of raw materials and logistics
Storing
Safety management

For applications in the following sectors
Car industry
Aviation industry
Shipping industry
Railway industry
Construction and building industry
Fastening technology
Electronics
Medicine and hygiene
Labels, adhesive tapes, packaging
Textiles
System providers

EMS-CHEMIE AG
Business Unit EMS-GRILTECH
Via Innovativa 1
7013 Domat/Ems
Phone +41 81 632 72 02
Fax +41 81 632 74 02
Email: info@emsgriltech.com
www.emsgriltech.com

Member of FKS

Company

Year of formation
1936 founded as Holzverzuckerungs AG (HOVAG),
1960 renamed into EMSER WERKE AG and finally into
EMS-CHEMIE AG in 1981.

Size of workforce
2,855 worldwide in December 2015

Sales channels
Direct Sales Channel, Distributors/Traders

Contact partners
Application technology and sales:
Phone: +41 81 632 72 02, Fax: +41 81 632 74 02
Email: info@emsgriltech.com, www.emsgriltech.com

Contact partners
The business unit EMS-GRILTECH is part of EMS-CHE-
MIE AG which belongs the EMS-CHEMIE HOLDING AG.
We manufacture and sell Grilon, Nexylon and Nexylene
fibers, Griltex hotmelt adhesives, Grilbond adhesion
promoters, Primid crosslinkers for powder-coatings and
Grilonit reactive diluents. We have developed these
materials and additives into excellent specialty products
for technically demanding applications. In this way we
create added value for our customers as they can also
only be successful with continual improvements.

Thermoplastic hotmelt adhesives
Thermoplastic adhesive products for technical and textile bonding applications are sold under the trade name
„Griltex®". EMS-GRILTECH has many years of experience
in the manufacture of tailor-made copolyamides and
copolyesters for different application fields. The melt
temperatures and melt viscosity can be modified over
a wide range depending on the different requirements
of each application. The adhesives are available as
powder in a wide range of grain sizes or as granules.
Manufacturing is carried out in our own polymerisation
and grinding plants.

Griltex® ES – Bonding of Even Surfaces
Hotmelt adhesives for bonding of metal, plastics, glass
and other smooth surfaces are produced under the trade
name Griltex® ES.

Range of Products

Types of adhesives
Hot melt adhesives

Types of sealants
Other

Raw materials
Additives
Resins
Polymers

Equipment, plant and components
for conveying, mixing, metering and for adhesive application
measuring and testing

For applications in the field of
Paper/packaging
Wood/furniture industry
Construction industry, including floors, walls and ceilings
Electronics
Mechanical engineering and equipment construction
Automotive industry, aviation industry
Textile industry
Adhesive tapes, labels
Hygiene
Household, recreation and office

Griltex® CE/CT for Composite Applications
For fiber-based composites made of glass-, basalt-,
carbon- and polymer-fibers EMS produces tailor-made
materials. These are established under the trade names
Griltex® CE and Griltex® CT.

EMS-GRILTECH's corporate offices with research
laboratory, technical service centre and production
facility are located at Domat/Ems, Switzerland. We also
have further production plants and technical service
centers located at Sumter, SC (USA) and Neumünster
(Germany). In Japan and Taiwan we have sales offices
and customer service labs.

EMS-GRILTECH is present worldwide either with its own
sales companies or represented by agents.

EUKALIN Spezial-Klebstoff Fabrik GmbH

Ernst-Abbe-Straße 10
D-52249 Eschweiler
Phone +49 (0) 24 03-64 50-0
Fax +49 (0) 24 03-64 50-26
Email: eukalin@eukalin.de
www.eukalin.com

Member of IVK

Company

Year of formation
1904

Size of workforce
60 employees

Managing partners
Dr. Joachim Schulz
Timm Koepchen

Ownership structure
family owned

Subsidiaries
EUKALIN Corp. USA

Sales channels
Directly and through dealers

Contact partners
Application technology and sales:
Timm Koepchen

Range of Products

Types of adhesives
Hot melt adhesives
Dispersion adhesives
Dextrin and starch adhesives
Glutine glue
Pressure-sensitive adhesives
Latex adhesives
Cold seal adhesives
Polyurethane Adhesives

Equipment, plant and components
for conveying, mixing, and metering
for adhesive application

For applications in the field of
Paper/packaging
Bookbinding/graphic design
Textile industry
Flexible packaging
Building industry
Tapes + Labels
Labelling

Evonik Industries AG
D-45764 Marl, www.evonik.com/crosslinkers,
www.evonik.com/adhesives-sealants,
www.evonik.com/designed-polymers
D-45764 Marl, www.vestamelt.de
D-64293 Darmstadt, www.visiomer.com
D-45127 Essen, www.evonik.com/polymer-dispersions
www.evonik.com/hanse, www.evonik.com/tegopac
D-63457 Hanau, www.aerosil.com,
www.dynasylan.com, www.evonik.com/fp

Member of IVK

Company

Year of formation
2007

Ownership structure
74.9 % RAG Stiftung, 25.1 % CVC

Contact partners
Application technology and sales:

Resource Efficiency
Phone +49 (0) 76 23-91-83 92
(application techn.)
Phone +49 (0) 61 81-59-34 76 (sales)
Email: aerosil@evonik.com, fillers.pigments@evonik.com
Phone +49 (0) 23 65-49-48 43
Fax +49 (0) 23 65-49-50 30
Email: adhesives@evonik.com
Phone +49 (0) 23 65-49-43 56 (VESTAMELT®)
　　　 +49 (0) 61 51-18-10 02 (VISIOMER®)
Email: vestamelt@evonik.com, visiomer@evonik.com

Nutrition & Care
Phone +49 (0) 2 01-1 73 21 33 (Sales)
Email: info@polymerdispersion.com, hanse@evonik.com, TechService-Tegopac@evonik.com

Further information
Evonik, the creative industrial group from Germany, is one of the world leaders in specialty chemicals. Profitable growth and a sustained increase in the value of the company form the heart of Evonik's corporate strategy. Its activities focus on the key megatrends health, nutrition, resource efficiency and globalization. Evonik benefits specifically from its innovative prowess and integrated technology platforms. Evonik is active in over 100 countries around the world with more than 35,000 employees. In fiscal 2016 the enterprise generated sales of around € 12,7 billion and an operating profit (adjusted EBITDA) of about € 2.165 billion.

Range of Products

Types of adhesives
Hot melt adhesives (VESTAMELT®) (DYNACOLL® S)

Types of sealants
Acrylic sealants (DEGALAN®)

Raw materials
Additives: waxes (VESTOWAX®, SARAWAX®), defoamer (TEGO® Antifoam), wetting agents (TEGOPREN®), thickener (TEGO® Rheo), silica nanoparticles (Nanopox®), silicone rubber particles (Albidur®), methacrylate monomers (VISIOMER®), pyrogenic silicas and metal oxides (AEROSIL®, AEROXIDE®), specialty precipitated silicas (SIPERNAT®), functional silanes (Dynasylan®)

Crosslinkers: speciality resins, aliphatic diamines (VESTAMIN®), aliphatic isocyanates (VESTANAT®)

Polymers: amorphous poly-alpha-olefines (VESTOPLAST®), copolyesters (DYNACOLL®), liquid polybutadienes (POLYVEST®), polyacrylates (DEGALAN®, DYNACOLL® AC), silane-modifiied polymers (Polymer ST, TEGOPAC®), condensation curing silicones (Polymer OH)

For applications in the field of
Paper/packaging
Bookbinding/graphic design
Wood/furniture industry
Construction industry, including floors,
Walls and ceilings
Electronics
Automotive industry, aviation industry
Textile industry
Adhesive tapes, labels
Hygiene
Manufacturing of hotmelt adhesives
Wind energy

ExxonMobil Chemical Central Europe
A division of ESSO Deutschland GmbH

Neusser Landstraße 16 · D-50735 Köln
Phone +49 (0) 2 21-770-31 · Fax +49 (0) 2 21-770-33 20
www.exxonmobil.de

Member of IVK

Company	Range of Products

Contact partner
Heide Henseler
Phone +49 (0) 221-77 03-296
Email: heide.henseler@exxonmobil.com

Types of adhesives
Hot melt adhesives
Solvent-based adhesives

Types of sealants
Butyl sealants
Other

Raw materials
Resins
Solvents
Polymers

For applications in the field of
Paper/packaging
Bookbinding/graphic design
Wood/furniture industry
Construction industry, including floors,
walls and ceilings
Electronics
Mechanical engineering and equipment
construction
Automotive industry, aviation industry
Textile industry
Adhesive tapes, labels
Hygiene
Household, recreation and office

Fermit GmbH

Zur Heide 4
D-53560 Vettelschoß
Phone +49 (0) 26 45-22 07
Fax +49 (0) 26 45-31 13
Email: info@fermit.de
www.fermit.com

Member of IVK

Company

Year of formation
2008

Size of workforce
15

Ownership structure
100 % subsidiary of Barthélémy/France

Sales channels
sanitary and heating equipm. trade, industry, whole trade

Contact partners
Management:
Alois Hauk

Further information
former Nissen & Volk, Hamburg

Range of Products

Types of adhesives
Solvent-based adhesives
Resin-based adhesives

Types of sealants
Silicone sealants
MS/SMP sealants
Sealing pastes
Other

For applications in the field of
Sanitary and heating industry
Construction Industry
Mechanical engineering and equipment construction
Automotive industry, aviation industry

fischer Deutschland Vertriebs GmbH

Klaus-Fischer-Straße 1
D-72178 Waldachtal
Phone +49 (0) 74 43 12-0
Fax +49 (0) 74 43 12-42 22
www.fischer.de

Member of IVK

Company

Year of formation
1948

Size of workforce
4,600

Ownership structure
Privately owned

Subsidiaries
43 (AR, BE, BR, BG, CN, DK, DE, FI, FR, GR, IT, JP, HR, MX, NL, NO, AT, PL, RU, SG, SE, ES, KR, CZ, HU, US, UK, SK, PT, TU, AE, THA)

Sales channels
DIY, specialized trade

Contact partners
Management:
Ralf Häfele
Email: Ralf.Haefele@fischer.de

Further information
World market leader in chemical fastening systems

Range of Products

Types of adhesives
Reactive adhesives
Dispersion adhesives

Types of sealants
Acrylic sealants
Silicone sealants
MS/SMP sealants

For applications in the field of
Wood/furniture industry
Construction industry, including floors, walls and ceilings

Follmann GmbH & Co. KG

Heinrich-Follmann-Straße 1
D-32423 Minden
Phone +49 (0) 5 71-93 39-0
Fax +49 (0) 5 71-93 39- 300
Email: info@follmann.com
www.follmann.com

Member of IVK

Company

Year of formation
1977

Size of workforce
125 (2016)

Management
Dipl.-Chem. Dr. Jörg Seubert
Dipl.-Ing. Hendrik Balcke

Subsidiaries
OOO Follmann
(Moscow/Russian Federation)
Follmann (Shanghai) Trading Co., Ltd.
(Shanghai/China)
Follmann Chemia Polska sp.z.o.o.
(Poznań/Poland)

Contact partners
Head of Sales:
Maik Schmolke
Product Management:
Torsten Krite

Range of Products

Types of adhesives
Hot melt adhesives
Dispersion adhesives

Raw materials
Polymers

For applications in the field of
Paper/packaging
Bookbinding/graphic arts
Wood/furniture industry

Forbo Eurocol Deutschland GmbH

August-Röbling-Straße 2
D-99091 Erfurt
Phone +49 (0) 3 61-7 30 41-0
Fax +49 (0) 3 61-7 30 41-91
Email: info.erfurt@forbo.com
www.forbo-eurocol.de

Member of IVK, FCIO, VLK

Company

Year of formation
1920

Size of workforce
98

Managing partners
Forbo Beteiligungen GmbH,
D-79539 Lörrach

Nominal capital
2.050.000 EUR

Ownership structure
100 % share holder

Sales channels
direct sales and/or through agents

Contact partners
Management:
Ruediger Beez,
Jochen Schwemmle (Chairman)

Application technology and sales:
Ruediger Beez

Range of Products

Types of adhesives
Reactive adhesives
Solvent-based adhesives
Dispersion adhesives
Pressure-sensitive adhesives

For applications in the field of
Construction industry, including floors,
walls and ceilings

H.B. Fuller

Connecting what matters.™

H.B. Fuller Europe GmbH
Talacker 50
CH-8001 Zürich
www.hbfuller.com

Member of IVK, FKS, VLK

Company

Global reach
- H.B. Fuller operates three regional headquarters across the globe:
  Americas – St. Paul, Minn., U.S.
  EIMEA – Zurich, Switzerland
  Asia Pacific – Shanghai, China
- Direct presence in 35 countries and customers in more than 100 geographic markets.

European commitment
H.B. Fuller has a network of specialised production sites across Europe serving customers in electronics, disposable hygiene, medical, transportation, clean energy, packaging, construction, woodworking, general industries and other consumer businesses.

About H.B. Fuller
For 130 years, H.B. Fuller has been a leading global adhesives provider focusing on perfecting adhesives, sealants and other specialty chemical products to improve products and lives. With fiscal 2016 net revenue of $2.1 billion, H.B. Fuller's commitment to innovation brings together people, products and processes that answer and solve some of the world's biggest challenges. Our reliable, responsive service creates lasting, rewarding connections with customers. And our promise to our people connects them with opportunities to innovate and thrive. For more information, visit us at www.hbfuller.com and subscribe to our blog.

Range of Products

Our Technologies
- Hot Melt
- Polymer and Specialty Technologies
- Reactive Chemistries: Urethane Epoxy Solventless
- Solvent-based
- Water-based

Our markets
- Automotive
- Building and Construction
- Consumer Products
- Electronic and Assembly Materials
- Emulsion Polymers
- General Assembly
- Nonwovens and Hygiene
- Packaging
- Paper Converting
- Woodworking

Committed to our communities
- Supporting STEM education and youth leadership development
- Volunteers reach more than 30 countries with 5,000-plus hours of service annually
- Company and employee donations total more than $ 1.5 million globally each year.

GLUDAN
Deutschland GmbH

Am Hesterkamp 2
D-21514 Büchen
Phone +49 (0) 41 55-49 75-0
Fax +49 (0) 41 55-49 75 49
Email: gludan@gludan.de
www.gludan.com

Member of IVK

Company

Year of formation
1977

Size of workforce
28

Ownership structure
GmbH

Sales channels
Own sale force, and agents, have a look on
our map www.gludan.com for contacts

Contact partners
Management:
ks@gludan.de

Application technology and sales:
od@gludan.de
sb@gludan.de

Further information
www.gludan.com

Range of Products

Types of adhesives
Hot melt adhesives
Dispersion adhesives
Glutine glue
Pressure-sensitive adhesives

Raw materials
Additives
Fillers
Polymers
Starch

Equipment, plant and components
for conveying, mixing, metering and
for adhesive application
measuring and testing

For applications in the field of
Paper/packaging
Bookbinding/graphic design
Wood/furniture industry
Construction industry, including floors,
walls and ceilings
Textile industry
Adhesive tapes, labels
Hygiene
Household, recreation and office

Fritz Häcker GmbH + Co. KG

Im Holzgarten 18
D-71665 Vaihingen/Enz
Phone +49 (0) 70 42 - 94 62-0
Fax +49 (0) 70 42 - 9 89 05
Email: info@haecker-gel.de
www.haecker-gel.de

Member of IVK

Company

Year of formation
1885

Size of workforce
22

Ownership structure
private owned company

Sales channels
Distributers worldwide

Contact partners
Management:
Ralf Müller

Range of Products

Types of adhesives
Gelatine based Adhesives

Types of sealants
Other

Raw materials
Technical Gelatine

Equipment, plant and components
for conveying, mixing, metering and for
adhesive application
measuring and testing

For applications in the field of
Paper/packaging
Bookbinding/graphic art
Tissue and towels
Box covering and lamination

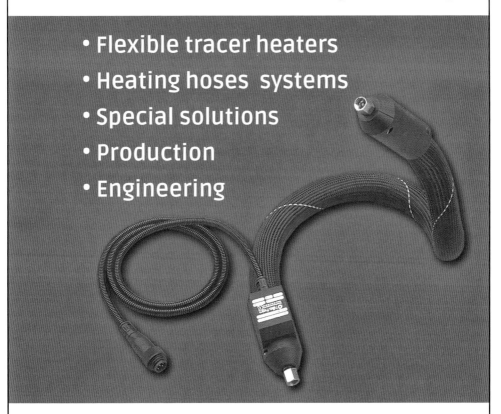

Henkel
AG & Co. KGaA
Henkelstraße 67
D-40191 Düsseldorf
Phone +49 (0) 211-797-0
www.henkel.com

Member of IVK, FCIO, FKS, VLK

Company

Ownership structure
AG & Co. KGaA

Contact partners
Business Unit Adhesive Technologies
Headquarters Düsseldorf:
Phone +49 (0) 211-797-0
www.henkel-adhesives.com

Henkel Central
Eastern Europe GmbH
Phone +43 (1) 7 11 04-0
www.henkel.at
www.henkel-cee.com

Henkel & Cie. AG
Phone +41 (61) 825-70 00
www.henkel.ch

Henkel Belgium N.V.
Phone +32 (2) 421-27 11
www.henkel.be

Further information
Henkel is the leading solution provider for
adhesives, sealants and functional coatings worldwide in two business segments:
Industry, and Consumers, Craftsmen and
Building. We offer a comprehensive portfolio of tailor-made solutions for our
customers and of high-quality consumer
products. Our global team of experts
partners with customers to deliver best-in-
class service. Our powerful innovations
and leading technologies under top-brands
such as Loctite create sustainabele value

Range of Products

Types of adhesives
Hot melt adhesives
Reactive adhesives
Solvent-based adhesives
Dispersion adhesives
Vegetable adhesives, dextrin and
starch adhesives
Pressure-sensitive adhesives
Instant adhesives
Structural adhesives
Anaerobic Adhesives
Laminating Adhesives

Types of sealants
Acrylic sealants
Butyl sealants
PUR sealants
Silicone sealants
MS/SMP sealants
Other

for industrial customers, consumers and
professional users. Our industrial product
portfolio is organized into five Technology
Cluster Brands – Loctite, Bonderite,
Technomelt, Teroson and Aquence. For
consumer and professional markets, we
focus on the four global brand platforms
Pritt, Loctite, Ceresit and Pattex.

IMCD Deutschland GmbH & Co. KG
Konrad-Adenauer-Ufer 41 – 45
D-50668 Köln
Phone +49 (0) 2 21-77 65-0
Fax +49 (0) 2 21-77 65-3 05
Email: coatings@imcdgroup.com
www.imcdgroup.com

Member of IVK

Company

Year of formation
1960

Size of workforce
135

Ownership structure
100 % IMCD N.V.

Sister Companies
IMCD operates all over Europe, Brazil,
Asia-Pacific and Africa

Contact partners
Management:
Piet van der Slikke, CEO IMCD N.V.
Frank Schneider, MD IMCD Deutschland

Application technology and sales:
Dr. Heinz-J. Küppers, Industry Manager
Adhesives

Further information
IMCD Deutschland is part of IMCD N.V.,
a leading international company in sales,
marketing and distribution of specialty
chemicals and food ingredients. With over
1,700 high-calibre professionals in more than
40 countries, IMCD offers a unique combination of local understanding backed by an
impressive international infrastructure. Listed
at Euronext, Amsterdam (IMCD.AS), IMCD
realised revenues of € 1,714 million in 2016.
IMCD Deutschland benefits from the affilia-
tion to this successful international network
as well as from the many years of experience
in the German market.

Range of Products

Types of adhesives
Solvent-based adhesives
Solvent-free adhesives
Dispersion adhesives
Pressure sensitive adhesives

Raw materials
Additives, Fillers, Resins, Solvents, Polymers

For applications in the field of
Paper/Packaging
Woodworking/Joinery
Building, Construction, Civil Engineering
Electronics
Automotive
Textile
Tapes, Labels, Graphic Arts

Jowat SE
Ernst-Hilker-Straße 10 – 14
D-32758 Detmold
Phone +49 (0) 52 31-7 49-0
Fax +49 (0) 52 31-7 49-1 05
Email: info@jowat.de
www.jowat.com

Member of IVK, FKS

Company

Year of formation
1919

Size of workforce
approx. 1,000

Managing partners
Klaus Kullmann
Ralf Nitschke
Dr. Christian Terfloth

Ownership structure
Shareholder company (not publicly traded)

Subsidiaries
20 worldwide

Sales channels
Own affiliations and distributors

Contact partners
Product marketing
Timm Schulze

Range of Products

Types of adhesives
Hot melt adhesives
Reactive adhesives
Solvent-based adhesives
Dispersion adhesives
Pressure-sensitive adhesives
Separating agents

For applications in the field of
Paper/packaging
Bookbinding/graphic design
Wood/furniture industry
Construction industry, including floors,
walls and ceilings
Electronics
Automotive industry, aviation industry
Textile industry
Adhesive tapes, labels
Upholstery, mattresses

KANEKA BELGIUM NV

The Dreamology Company
—Make your dreams come true—

Kaneka Belgium N. V.
MS Polymer Department
Nijverheidsstraat 16
B-2260 Westerlo (Oevel)
Telefon +32 (0) 14 - 25 45 20
Telefax +32 (0) 14 - 25 78 87
E-Mail: info.mspolymer@kaneka.be
www.kaneka.be

Member of IVK

Company

Year of formation
Kaneka Belgium N.V. was founded in 1970 as the European production site of the globally acting Kaneka Corporation, Japan.

Further information
Kaneka products have conquered the European market, becoming a synonym for premium quality raw materials, with the brand MS POLYMER basically defining a new group of adhesives and sealants. Other brands of Kaneka's MS Polymer Department are SILYL and XMAP.

Production/Technical Service contact
Kaneka Belgium N.V.
MS Polymer Department
Nijverheidsstraat 16
B-2260 Westerlo (Oevel)
Phone +32 14-25 78 67
Email: info.mspolymer@kaneka.de

Marketing contact
Kaneka Belgium N.V.
MS Polymer Department
Nijverheidsstraat 16
B-2260 Westerlo (Oevel)
Phone +32 14-25 45 20
Email: info.mspolymer@kaneka.de

For contact in D, A, CH, CEE
Werner Hollbeck GmbH
Kirchmannstraße 22
D-45133 Essen
Phone +49 (0) 2 01-7 22 16 16
Fax +49 (0) 2 01-7 22 16 06
Email: info@hollbeck.de

Range of Products

Raw materials
MS Polymer, SILYL and XMAP are reactive high performance polymers based on polyether, or polyacrylate, respectively. Customers can select between moisture curing grades, addition cure types, and radical cure types including UV-cure. The polymers' characteristics produce elastic adhesives and sealants, but also high strength types and coatings can be formulated.

End products' features
Solvent and isocyanate free 1-K/2-K-reactive adhesives and sealants
Pressure sensitive adhesives (PSA)
Oil resistance, temperature resistance (150 °C permanent use, XMAP)
Low gas and moisture permeability
High UV-resistance
Adhesive blends with epoxies

For applications in the field of
Wood/furniture industry
Construction industry, including floors, walls and ceilings
Electronics
Automotive industry, transportation industry
Adhesive tapes, labels
Household, recreation and office
General industry
Shipbuilding industry
Waterproofing and roof coatings

 **KEYSER & MACKAY**

Keyser & Mackay
German branch office
Industriestraße 163
D-50999 Köln (Rodenkirchen)
Phone +49 (0) 22 36-39 90-0
Fax +49 (0) 22 36-39 90-33
Email: info.de@keysermackay.com
www.keysermackay.com

Member of IVK

Company

Founded
1894

Size of workforce
120

Subsidiaries
Headquarter in The Netherlands

Subsidiaries in Germany, Belgium, France, Switzerland, Poland, Spain

Contact partners
Mr. Robert Woizenko
Email: r.woizenko@keymac.com

Types of adhesives
Hot melt adhesives
Reactive adhesives
Solvent-based adhesives
Dispersion adhesives
Pressure-sensitive adhesives

Types of sealants
Acrylic sealants
Butyl sealants
Polysulfide sealants
PUR sealants
Silicone sealants
MS/SMP sealants

Range of Products

Raw materials
Additives: Oxazolidines, adhesion promoter, flame retardants, thickener

Fillers: precipitated calcium carbonate, talc, carboxymethylcellulose, barium sulphate

Resins: hydrogenated hydrocarbon resins, C5/C9-hydrocarbon resins, pure monomer resins, modified rosin resins, resin dispersions

Polymers: APO, PE/PP waxes, STP, SIS/SBS/SEBS, acrylic copolymers, acrylic dispersions

For application in the field of
Paper/packaging
Bookbinding/graphic design
Wood/furniture industry
Construction industry, including floors, walls and ceilings
Electronics
Mechanical engineering and equipment construction
Automotive industry, aviation industry
Textile industry
Adhesive tapes, labels
Hygiene
Household, recreation and office

Kiesel Bauchemie GmbH u. Co. KG

Wolf-Hirth-Straße 2
D-73730 Esslingen
Phone +49 711 93134-0
Fax +49 711 93134-140
Email: kiesel@kiesel.com
www.kiesel.com

Member of IVK, FCIO

Company

Year of formation
1959

Size of workforce
160

Ownership structure
family owned

Subsidiaries
Sales offices in Benelux, Poland, Czech
Republic, Switzerland, France

Sales channels
wholesale to professional installers

Contact partners
Management & Sales:
Beatrice Kiesel-Luik

Application technology:
Ulrich Lauser

Range of Products

Types of adhesives
Reactive adhesives
SMP-Products
Silane-modified polymers
Dispersion adhesives

Types of sealants
Acrylic sealants
PUR sealants

For applications in the field of
Construction industry, including floors,
walls and ceilings

KISLING
Deutschland GmbH

Drillberg
D-97980 Mergentheim
Phone +49 (0) 791-407 27-0
Fax +49 (0) 791-407 27-50
Email: dinfo@kisling.com

Member of IVK, FKS

Company

Year of formation
2000

Size of workforce
5

Nominal capital
100,000 €

Ownership structure
100 % subsidiary of Kisling AG, Wetzikon

Sales channels
direkt marketing and distribution

Contact partners
Application technology and sales:
Heiko Haupt

Range of Products

Types of adhesives
Reactive adhesives

For applications in the field of
Wood/furniture industry
Electronics
Mechanical engineering and equipment
construction
Automotive industry, aviation industry
Household, recreation and office

KLEIBERIT Adhesives
KLEBCHEMIE M.G. Becker GmbH & Co. KG
Max-Becker-Straße 4
76356 Weingarten/Germany
Phone +49 7244 62-0
Fax +49 7244 700-0
Email: info@kleiberit.com
www.kleiberit.com

Member of IVK

Company

Year of formation
1948

Size of workforce
over 550 worldwide

Ownership structure
Owner Managed GmbH & Co. KG

Subsidiaries
Australia, France, USA, Canada, UK, Japan, China, Singapore, Russia, Brazil, India, Mexico, Ukraine, Turkey, Belarus

Sales channels
Direct, Wholesale

Contact partners
Management:
Dipl. Phys. Klaus Becker-Weimann
Dr. Achim Hübener

Sales:
Wolfgang Hormuth

Further information
Specialist in PUR-Adhesive-Technology
Competence PUR

Range of Products

Types of adhesives
PUR Hot melt
Reactive Hot melt (PUR, POR)
Holt melt (EVA, PO, PA)
1-C and 2-C reactive adhesives
(PUR, STP, Epoxy)
PUR foam systems
Dispersion adhesives
(Acrylat, EVA, PUR, PVAC)
Sealants and assembly adhesives
Pressure-sensitive adhesive
EPI-systems
Solvent-based adhesives

Coating systems
Kleiberit HotCoating® based on PUR
TopCoating based on UV lacquer

For applications in the field of
Wood- and furniture industry
Profile wrapping
Construction industry including floors, walls, ceilings and façade elements
Sandwich Panels
Textile industry
Automotive industry
Ship and boat building
Bookbinding industry
Surface finishing
Doors, windows, stairs and flooring
Filtration industry
Paper and packaging industry

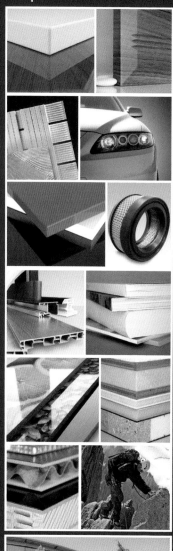

KÖMMERLING CHEMISCHE FABRIK GMBH

**Kömmerling
Chemische Fabrik GmbH**
Zweibrücker Straße 200
D-66954 Pirmasens
Phone +49 (0) 63 31 56-20 00
Fax +49 (0) 63 31 56-19 99
Email: info@koe-chemie.de
www.koe-chemie.de

Member of IVK

Company

Year of formation
1897

Size of workforce
450

Ownership structure
Royal Adhesives & Sealants

Subsidiaries
Kömmerling Chimie SARL, Strasbourg (F)
Kommerling UK LTD, Uxbridge (UK)
Kömmerling chemische Fabrik GmbH,
Beijing (China)

Sales channels
B2B, Trading

Contact partners
Management:
Bernd Helfrich
C. Richard Spalton

Technology and Sales:
Head of Sales Industrial Adhesives:
Dr. Gert Heckmann;
Head of Sales Glass:
Herbert Haas
Head of Technology:
Dr. Norbert Schott

Further information
Kömmerling Chemische Fabrik GmbH is a
leading international manufacturer of high
quality adhesives and sealants. Established
over 115 years ago, Kömmerling is today a
major systems supplier for the Glass,
Transport, Construction, Industrial Assembly
and Renewable Energy industries.

Range of Products

Types of adhesives
Hot melt adhesives
Reactive adhesives
Solvent-based adhesives
Dispersion adhesives
Pressure-sensitive adhesives

Types of sealants
Butyl sealants
Polysulfide sealants
PUR sealants
Silicone sealants
MS/hybrid sealants

For applications in the field of
Construction
Automotive
Marine
Coil coating
Shoe industry
Insulating glass
Photovoltaic, Solar Thermal
Window Bonding
Structural glazing
Wind energy

Krahn Chemie GmbH
Grimm 10
D-20457 Hamburg
Phone +49 (0) 40-3 20 92-0
Fax +49 (0) 40-3 20 92-3 22
Email: info@krahn.de
www.krahn.de

Member of IVK

Company

Krahn Chemie is a speciality chemical distribution company and represents renowned, globally operating manufacturers in Europe. One of our core segments is the adhesives & sealants industry to which we provide raw materials.

Year of formation
1972

Size of workforce
160

Managing directors
Dr. Rolf Kuropka
Axel Sebbesse

Ownership structure
Otto Krahn (GmbH & Co.) KG founded 1909

Subsidiaries
Krahn Chemie Polska Sp. z o.o.
Krahn Chemie Benelux BV
Pietro Carini S.p.A.

Contact partners
Business Segment Manager
Adhesives & Sealants
Thorben Liebrecht
Email: thorben.liebrecht@krahn.eu

Application Technology and Sales:
Dr. Hartmut Salow,
Email: hartmut.salow@krahn.eu

Supplying partners
Baerlocher, Bostik, Celanese, Dixie Chemical Company, Eastman Chemical, ExxonMobil Chemical, Galstaff Multiresine, Gulf Chemical International, Lord Germany, Otsuka Chemical, Oxea, Sinopec, Tosoh, Synegis BVBA, Epaflex, Elachem, Polytex, Nanjing SiSiB Silicones, Higree, UV Chem-Keys, Dynaplak Adhesives & Starches, Chromaflo

Range of Products

Raw materials
Additives:
Pigments, Pigment Pastes
Silanes, Photoinitiators
Light Stabilizers, Plasticizers
Cellulose Esters, Cellulose Ethers

Dispersions:
Acrylic Dispersions
PVAc Dispersions
VAE Dispersions
Polychloroprene Latex
PVB Dispersions
Biobased Dispersions

Resins and Hardeners:
Hydrocarbon Resins
Saturated Polyester Polyols
Saturated Polyester Resins
Unsaturated Polyester Resins
PU-Prepolymers
UV curing Resins
Isocyanates
Epoxy Hardeners

Polymers:
CR, CSM, EVA, TPU,
SIS, SEBS

Adhesion Promoters

For applications in the field of
Paper/packaging
Bookbinding/graphic design
Wood/furniture industry
Construction industry, including floors, walls and ceilings
Electronics
Traffic Industry, Automotive industry, aviation industry
Textile industry
Adhesive tapes, labels
Mechanical engineering and equipment construction DIY

Kraton Polymers Nederland B.V.
Transistorstraat 16
NL-1322CE Almere
Phone +31 36 5462 800
Email: info@kraton.com

Member of IVK

Company

Contact details
info@kraton.com

Sustainable Solutions.
Endless Innovation.
At Kraton, we deliver exceptional value to our customers through compelling, innovative and sustainable solutions. We create, develop and manufacture renewable chemicals and specialty polymers to meet our customers' performance and market needs. We contribute to a sustainable future by combining superior product performance and quality, with reliability of global supply. We live our core values every day, including taking ownership for doing what is safe, honest and ethical. At Kraton, we seek solutions that are sustainable, and believe innovation is endless.

Range of Products

Types of adhesives
Hot melt adhesives
Solvent-based adhesives
Pressure-sensitive adhesives

Raw materials
Polymers: Styrenic block copolymers, Polyisoprene-rubber

Pine Chemical Tackifiers

For applications in the field of
Paper/packaging
Bookbinding/graphic design
Wood/furniture industry
Construction industry, including floors, walls and ceilings
Automotive industry, aviation industry
Textile industry
Adhesive tapes, labels
Hygiene

LANXESS Deutschland GmbH

Kennedyplatz 1
D-50569 Cologne
Phone +49 (0) 221-8885-0
www.lanxess.com

Member of IVK

Company

The global chemical company LANXESS with head office in Cologne, Germany, has been listed at the German stock exchange since 2005. More than 19,200 employees are working for LANXESS worldwide.

The product portfolio consists of 5 segments:
- Advanced Intermediates
- Specialty Additives
- Engineering Materials
- Performance Chemicals
- ARLANXEO*

The turnover was 7.7 bn € in 2016.

*ARLANXEO was established in April 2016 as a joint venture of LANXESS and Saudi Aramco. The company's core business is the development, manufacturing and marketing of synthetic high-performance rubber for use in the automotive and tire industries, the construction industry, and the oil and gas industries.

Contact
For Baypren® ALX and Levamelt®
Dr. Martin Schneider
ARLANXEO Deutschland GmbH
High Performance Elastomers
Phone: +49 (0) 221 8885 5908
Email: martin.schneider@arlanxeo.com

For biocides:
Dr. Peter Wachtler
Material Protection Products
Technical Marketing – Industrial Preservation
Phone: +49 (0) 221 8885 4802
Email: peter.wachtler@lanxess.com

Range of Products

Raw materials and additives for the production of adhesives and sealants:

Levamelt® and Baypren® ALX
Their variety of polarity, viscosity and crystallization rate make the synthetic rubber from the Levamelt® and Baypren® ALX product series a perfect fit for different applications in adhesives production.

Biocides
Wide range of biocides under the brand names Preventol® and Metasol®:
- In-can preservatives for all kinds of water-based adhesives
- Fungicides for film protection of sealants, adhesives and grout fillings
- Consultation on microbiological questions

Lohmann GmbH & Co. KG
Irlicher Straße 55
D-56567 Neuwied
Phone +49 (0) 26 31-34-0
Fax +49 (0) 26 31-34-66 61
Email: info@lohmann-tapes.com
www.lohmann-tapes.com

Member of IVK

Company

Year of formation
1851

Size of workforce
approx. 1,700 worldwide

Managing directors:
Elmar Boeke
Martin Schilcher

Subsidiaries
I, F, E, PL, A, GB, NL, DK, SE, RU, UA, USA, China, Korea, Singapore, India and Mexico

Sales channels
Market segments:
Graphics, Consumer Goods, Building & Construction, Transportation, Mobile Communication, Medical Technology, Renewable Energies, Electronics and Hygiene

Contact partners
Head of PR and Corporate Communication
Christina Barg-Becker M. A.

Further information
Lohmann offers mainly customized bonding solutions and takes care of its customers from the first idea up to automatic applications.
This is also shown in the company logo: „The Bonding Engineers".

Range of Products

Types of adhesives
Hot melt adhesives
Reactive adhesives
Solvent-based adhesives
Dispersion adhesives
Pressure-sensitive adhesives

Types of sealants
Acrylic sealants

Raw materials
Polymers

Equipment, plant and components for
Conveying, mixing, metering
Adhesive application
Surface pretreatment
Adhesive curing and drying
Measuring and testing

For applications in the field of
Paper/packaging
Plate mountig tapes
Wood/furniture industry
Building & Construction industry, including floors, walls and ceilings, structural glazing
Electronics
Automotive industry
Adhesive tapes, die-cuts, labels
Mobile Communication
Renewable Energies
Medical Technology/diagnostics
Hygiene
Consumer Goods
High security cards

Diversity that pays off

...ecause bonding is the smarter joining technology for many industries.

...o matter what complex challenges you may face today or in future: With Lohmann you will bond even ...ore reliably, efficiently and economically. The Bonding Engineers analyze your demands, translate the ...esired application into the matching adhesive technology and integrate it into your process. On the ...asis of a diverse product portfolio. And with accompanied service that ensures long-lasting success.

...e at Lohmann call this philosophy "Smart Bonding Approach". You can call it the perfect bonding solution ...r your individual application.

...r further information please visit us at:
...ww.lohmann-tapes.com

The Bonding Engineers

LORD Germany GmbH
Itterpark 8
D-40724 Hilden
Phone +49 (0) 21 03-2 5 23 10
Fax +49 (0) 21 03-2 52 31 97
Email: monika_schulz@lord.com
www.lord.com/emea

Member of IVK

Company

Founded in
1924 by Hugh C. Lord

Employees
3,000 (LORD Corp. worldwide)

European Technical Center Hilden

Managing Directors
Dr. Dirk L. Schröder, Kristopher James Burson

Technical Service Contacts in Hilden
LORD Rubber to Metal adhesives
Dr. René Luther
Email: rene_luther@lord.com

LORD Automotive OEM 2K adhesives
Dipl. Ing. Marcus Lämmer
Email: marcus_laemmer@lord.com

LORD Industrial adhesives
Dipl. Ing. Marcus Lämmer
Email: marcus_laemmer@lord.com

LORD Electronic materials
Dipl. Ing. Marcus Lämmer
Email: marcus_laemmer@lord.com

LORD Flock adhesives and rubber coatings
Mrs. Dr. Christiane Stingel
Email: christiane_stingel@lord.com

Additional Information
LORD Corp. is the global leader in producing rubber to metal adhesives used in a wide variety of automotive, aerospace and industrial applications. In addition LORD is producing solvent based and aqueous flock adhesives and coatings for elastomeric components. On the structural assembly part LORD is producing 2K metal and composite structural adhesives

Range of Products

Type of Adhesives
2K Structural Adhesives (LORD®, FUSOR®, VERSILOK®)
Solvent based rubber to metal adhesives (CHEMLOK®, CHEMOSIL®)
Solvent based reactive 1K adhesives (LORD®, CHEMLOK®, FLOCKSIL®)
Water based reactive 1K adhesives (CUVERTIN®, SIPIOL®)
Solvent based aerospace coatings and primer (AEROGLAZE®)

For applications
Electronics: potting, coating, waver adhesives
Industrial: metal, composite, plastic
Elastomeric/Rubber: Rubber to metal, flock, slip-coating, anti-squeeze
Automotive (OEM): 2K cold cure hem flange adhesives, 2K LED cure hem flange sealer (body shop applied), 2K repair adhesives, vibration mounts, magnet rheological fluids (MR fluids) and MR devices
Aerospace: 2K OEM structural adhesives, 2K repair adhesives, coatings, vibration mounts, active vibration control devices

which are use e.g. in automotive hem flange bonding. Another focus is on electronic adhesives used for potting and encapsulation as well as electrically conductive adhesives and grease for thermal management.

The European Customer Service Center in Hilden is the research and development center for customized solutions as well as the main European training center.

LUGATO
GmbH & Co. KG

Großer Kamp 1
D-22885 Barsbüttel
Phone +49 (0) 40- 6 94 07- 0
Fax +49 (0) 40- 6 94 07- 1 10
Email: info@lugato.de
www.lugato.de

Member of IVK, FCIO,
Deutsche Bauchemie e.V.

Company

Year of formation
1919

Size of workforce
About 135

Ownership structure
GmbH & Co. KG, associated company of the
Ardex GmbH since 2005

Sales channels
DIY stores

Contact partners
Application technology and sales:
info@lugato.de

Further information
www.lugato.de
www.lugato.com

Range of Products

Types of adhesives
Dispersion adhesives
Cementitious adhesives
MS/SMP adhesives

Types of sealants
Acrylic sealants
Silicone sealants
MS/SMP sealants

For applications in the field of
Household, recreation and office

Mapei Austria GmbH

Fräuleinmühle 2
A-3134 Nußdorf o.d. Traisen
Phone +43 (0) 27 83 - 88 91
Fax +43 (0) 27 83 - 88 91 - 1 25
Email: office@mapei.at
www.mapei.at

Member of IVK, FCIO

Company

Year of formation
1980

Size of workforce
140 employees

Managing partners
Mapei SpA, Milano, Italien

Subsidiaries
Mapei Kft, Ungarn, Mapei sro,
Tschechische Republik;
Mapefin Austria GmbH

Contact partners
Management:
Mag. Andreas Wolf

Sales management:
Paul Solczykiewicz

Product management:
Ing. Reinhold Stinzl

Range of Products

Types of adhesives
Hot melt adhesives
Reactive adhesives
Solvent-based adhesives
Dispersion adhesives

Types of sealants
Acrylic sealants
Silicone sealants

For applications in the field of
Construction industry, including floors,
walls and ceilings

MAPEI S.p.A.

Via Catiero 22
20158 Milano, Italy
Phone +39 02/376 731
Fax +39 02/376 732 14
Email: mapei@mapei.it
www.mapei.com

Member of IVK

Company

Year of formation
1937

Size of workforce
more than 7,500

Managing partners
MAPEI S.p.A., Milano, Italy

Ownership structure
family-owned enterprise

Subsidiaries
81 subsidiaries in 32 countries

Sales channels
wholesalers

Contact partners
Management:
Flavio Terruzzi

Range of Products

Types of adhesives
Reactive adhesives
Dispersion adhesives
Pressure-sensitive adhesives

Types of sealants
Acrylics sealants
PUR sealants
Silicone sealants
MS/SMP sealants
Other

For applications in the field of
Construction industry, including floors, walls and ceilings

merz+benteli ag

more than bonding

Merbenit Gomastit Merbenature

merz+benteli ag
Freiburgstrasse 616
CH-3172 Niederwangen
Phone +41 (31) 980 48 48
Fax +41 (31) 980 48 49
Email: info@merz-benteli.ch
www.merz-benteli.ch

Member of FKS

Company

Organisation
merz+benteli ag was founded in 1918 to supply the Swiss watch industry with innovative adhesives. The enterprise is a family-owned joint-stock company specialised in adhesives and sealants. R&D and production in Switzerland.

Contact
Simon Bienz
Director Marketing & Sales

Distribution partners
DE/AT: Reiss Kraft GmbH
www.reiss-kraft.de / +49 7253 93 47 65

GR: Dialinas AE
www.dialinas.gr / +30 210 27 13 333

ES, PT, GB, IE, RU, DK, TR, SE, NO, FI, IT, BE, NL, LU, FR, CZ, SI
Distribution Groupe Europe
www.dge-europe.com / +31 172 436 361

IT: SKAB AG
www.skabag.com / +41 91 610 10 30

HU: Güteber Kft.
probond@berenyizoltan.hu / +36 1 213 5005

SL: Koop KOOP Trgovina d.o.o.
www.koop.si / +386 (7) 477-8820

IL: Rotal Adhesives & Chemicals Ltd.
www.rotal.com / +972 9 766 7990

USA, CAN, MEX: Chenso Inc.
www.chenso.com / +1 336 681 4131

Asia/Pacific: Innosolv Pty.
rick@innosolv.com / +61 (6) 9846 7871

Range of Products

Range of Products
SMP adhesives and sealants branded as GOMASTIT and MERBENIT.
Bio-based SMP sealants MERBENATURE
Patented MERBENTECH technology

Range of application
Elastic sealing for building construction:
Facades
Interior construction
Floor
Sanitary
Glazing
Roof
Fire protection

Elastic adhesives for industrial constructions:
Industry
Automotive
Marine

Michelman Deutschland GmbH
Alte Grenzstrasse 153
D-45663 Recklinghausen
Phone +49 (0) 23 61-66 05-0
Fax +49 (0) 23 61 - 66 05 - 55
Email: info@michelman.com
www.michelman.com

Member of IVK

Company

Year of formation
1949
(Michelman Inc. in Cincinnati, OH, USA)

Size of workforce
> 400 worldwide

Managing partners
S. J. Shifman, J. Rodgers, P. A. J. Griffith,
R. I. Michelman, D. M. Braun

Nominal capital
n/a

Ownership structure
Family owned

Subsidiaries
USA, Brazil, Belgium, Germany, India, China,
Singapore, Japan

Sales channels
Regional Sales Offices and global distributor
network

Contact partners
Marketing & Applications:
Dr. Volker Erb (Industry Manager),
volkererb@michelman.com

Sales:
Ulrich Balke (Regional Sales Manager EMEA),
ulrichbalke@michelman.com

Further information
Michelman products enhance the surface
qualities of coatings and inks, make composites
tougher, adhesives special and expand the possibilities of paper and film with barrier, functional,
and aesthetic features. Our extraordinary local
technical support helps ensure those products
deliver all they can to your bottom line.

Michelman provides ...
• Ready-made and custom formulations
• Leading-edge sustainable technology

Range of Products

Types of adhesives
Dispersion adhesives

Raw materials
Additives
Polymers

For applications in the field of
Paper/packaging
Bookbinding/graphic design
Wood/furniture industry
Construction industry, including floors,
walls and ceilings
Electronics
Mechanical engineering and equipment
construction
Automotive industry, aviation industry
Textile industry
Adhesive tapes, labels
Hygiene
Household, recreation and office

• Technical support for better processing and
performance
• Global resources to reach new markets
• Contract manufacturing to expand your
capabilities

Ecrylic™, Ecrovin™, Ecrothan™, ProHere™ and
Michem® technologies offer optimum protection, durability, and surface qualities in a wide
array of adhesives and for factory-applied and
field-applied special coatings on metal, concrete,
wood or plastic. Michem® additives add critical
qualities like block, rub, and water resistance,
and Licomer® system of polymer dispersions
and wax emulsions are state -of-the-art for the
care and maintenance of high traffic surfaces
in flooring, automotive, furniture, and soft goods
applications.

Minova CarboTech GmbH
Am Technologiepark 1
D-45307 Essen
Phone +49 (0) 201 80983 500
Fax +49 (0) 201 80983 9 500
Email: info.de@orica.com
www.minovaglobal.com

Member of IVK

Company

Holding company
Orica Ltd, 1st Nicholson Street,
3000 Melbourne, NSW, Australien

ORICA Subsidiaries worldwide
Australia, Chile, China, Germany, France,
Great Britain, India, Italy, Canada,
Kazakhstan, Austria, Poland, Romania,
Russia, Sweden, Switzerland, Singapore,
Spain, South Africa, Taiwan,
Czech Republic, Turkey, Ukraine, USA

Contact partners
Management:
Michael J. Napoletano, Andreas Humann

Application technology and sales
Udo Nielbock,
Phone +49 (0) 201 80983 721,
Mobil +49 (0) 172 266 3741,
Email: udo.nielbock@minovaglobal.com

Further information

Minova has more than 40 years of worldwide experience in the fields of water
control, ground stabilisation, anchoring,
rock bolting and flooring adhesives.

We are a leading supplier of products and
services for mining, tunnelling, geotechni-
cal engineering, special underground civil
engineering and floor technology.

Range of Products

Types of adhesives
Reactive adhesives

For applications in the field of
Wood/furniture industry,
Construction industry, including floors,
walls and ceilings

We supply an extensive range of products for:

- Anchoring systems for mining and tunnelling as well as special underground civil
  engineering
- Resin systems for sealing, rock stabilisation and reinforcement
- Accessories for injection and anchoring
  systems
- Adhesives for various floors and floor
  coverings

MÜNZING CHEMIE GmbH
Münzingstraße 2
D-74232 Abstatt
Phone +49 (0) 71 31-987-0
Fax +49 (0) 71 31-987-202
Email: sales.pca@munzing.com
www.munzing.com

Company

Year of formation
1830

Size of workforce
330

Managing partners
family owned

Subsidiaries
MÜNZING North America,
Bloomfield, NJ, USA
MÜNZING CHEMIE Iberia S.A.U.,
Barcelona, Spain
MÜNZING International S.a.r.l., Luxembourg
MÜNZING Micro Technologies GmbH,
Elsteraue, Germany
MÜNZING Shanghai Co. Ltd, P.R. China
MAGRABAR LLC, Morton Grove, IL, USA

Sales channels
direct and via distributors

Contact partners
Application technology:
Peter Bissinger
Tel. No. +49 (0) 71 31 - 987 - 174
Email: p.bissinger@munzing.com

Sales:
Dr. Nicholas Büthe
Tel. No. +49 (0) 71 31 - 987 - 148
Email: sales.pca@munzing.com

Range of Products

Raw materials
Additives: Defoamers, dispersants, rheology
modifiers, wetting and levelling agents

For applications in the field of
Paper/packaging
Bookbinding/graphic design
Wood/furniture industry
Construction industry, including floors, walls
and ceilings
Electronics
Mechanical engineering and equipment
construction
Automotive industry, aviation industry
Textile industry
Adhesive tapes, labels
Household, recreation and office

MUREXIN GmbH

Franz v. Furtenbach Straße 1
A-2700 WR. Neustadt
Phone +43 (0) 26 22/2 74 01
Email info@murexin.com
www.murexin.com

Member of IVK, FCIO

Company

Year of formation
1931

Size of workforce
400

Ownership structure
Schmid Industrie Holding

Subsidiaries
Czech. Republik, Hungary, Slovenia,
Russia, Slovak. Republic, Poland, France

Contact partners
Management:
Bernhard Mucherl

Range of Products

Types of adhesives
Solvent-based adhesives
Dispersion adhesives
MS adhesives

Types of sealants
Bitumen sealings
Bitumen-free sealings
Joint sealings

For applications in the field of
Construction industry, including floors,
walls and ceilings

NAGASE (EUROPA) GmbH

Immermannstraße 65 c
D-40210 Düsseldorf
Phone +49 (0) 2 11-8 66 20-0
Fax +49 (0) 2 11-3 23 70 68
Email: Service@Nagase.de
www.Nagase.de

Member of IVK

Company

Year of formation
1980

Size of workforce
31

Managing partners
Nagase & Co. Ltd. Tokyo

Nominal capital
1,200,000 €

Ownership structure
GmbH

Sales channels
Offices in London, Budapest

Contact partners
Management:
Mitsuru Kanno

Application technology and sales:
Heinrich Lüdeke

Further information
fine Chemicals, Pharmaceuticals, Cosmetics,
Enzymes, Plastics

Range of Products

Types of adhesives
Hot melt adhesives
Reactive adhesives
Solvent-based adhesives

Types of sealants
Acrylic sealants
Other

Raw materials
Additives
Resins
Solvents

For applications in the field of
Paper/packaging
Electronics
Automotive industry, aviation industry

Nordmann, Rassmann GmbH
Kajen 2
D-20459 Hamburg
Phone +49 (0) 40 36 87-0
Fax +49 (0) 40 36 87-2 49
Email: info@nrc.de
Internet: www.nrc.de

Member of IVK

Company

Year of formation
1912

Size of workforce
330

Managing Board
Dr. Gerd Bergmann, Irina Zschaler,
Carsten Güntner

Ownership structure
Family-owned

Subsidiaries
Austria, Bulgaria, Czech Republic, France, Hungary,
Italy, Poland, Portugal, Romania, Serbia, Slovac
Republic, Slovenia, Sweden, Switzerland, Turkey

Sales channels
Face-to-face sales, telephone, email, fax

Contact partners
Management: Henning Schild
Phone +49 (0) 40 36 87-248
Fax +49 (0) 40 36 87-72 48
Email: henning.schild@nrc.de

Application technology and sales: Michael Herrmann
Phone +49 (0) 40 36 87-463
Fax +49 (0) 40 36 87-74 63
Email: michael.herrmann@nrc.de

Further information
Founded in Germany in 1912, The NRC Group has
grown to become one of the world's leading sales
and marketing organizations in chemical distribution.

The NRC Group works in major industries such as
cosmetics, pharmaceuticals, food, construction, composites, coatings, adhesives, plastics, elastomers,
oleochemicals and lubricants to provide the keys to
success for both its customers and the chemical and
natural raw material manufacturers that it represents
from all around the globe, including:

- an extensive portfolio of products
- full market coverage
- technical and application expertise
- tailor-made supply chain solutions
- the support of self- or partner-owned laboratories, allowing for innovative product development

Range of Products

Raw materials
Additives: antioxidants/stabilizers, cellulose ethers,
dispersion powders, defoamers, PVA, polyethylene
oxide, rheology modifiers, polyolefin waxes and
copolymers (PE/PP), pigments, impact modifiers,
flame retardants and synergists, carbon fibers,
functional fillers, starch ethers, surfactants

Polymers: styrenic block copolymers (SBS, SEBS,
SEP, SIS, SIBS), isoprene rubber (IR), chloroprene
rubber (CR), PVDC

Resins: hydrogenated and non-hydrogenated
hydrocarbon resins, gum rosin and rosin ester deri-
vates epoxy resins and hardeners, phenolic resins,
alkyd resins

Reactive components: polyetherpolyols, polyesterpolyols, PTMEG, polycaprolactones, isocyanates
(MDI), catalysts, reactive diluents, chain extenders

Plasticizers: mineral oil based process oils,
gas-to-liquid (GtL) based process oils, DOTP, DPHP

Dispersions: styrene-acrylate, styrene-butadiene,
vinylacetate, chloroprene, isoprene, gum rosin ester,
Polyurethane

For applications in the field of
Adhesive tapes, labels
Automotive industry, aviation industry
Bookbinding/graphic design
Construction industry, including floors,
walls and ceilings
Electronics
Household, recreation and office
Hygiene
Mechanical engineering and equipment
construction
Paper/packaging
Textile industry
Wood/furniture industry

As an independent, family-owned business, now
with 16 European offices and over 100 years of
experience, The NRC Group and its staff of 330
employees generated a turnover of € 350 million
in 2016. For more details, please visit us online at
www.nrc-group.com

Nynas GmbH
Marktplatz 2
D-40764 Langenfeld
Email: thorsten.wolff@nynas.com
www.nynas.com

Member of IVK

Company

Year of formation
1983

Size of workforce
10

Managing partners
Nynas AB Stockholm

Nominal capital
76,693

Subsidiaries
30 offices worldwide

Contact partners
Sales Manager Germany:
Thorsten Wolff

Application technology and sales:
Nina Lulsdorf
Jan-Peter Laabs

Range of Products

Process oils for adhesives/hotmelts

For applications in the field of
Paper/packaging
Construction industry, including floors, walls
and ceilings
Automotive industry, aviation industry
Textile industry
Adhesive tapes, labels
Hygiene
Household, recreation and office

Omya GmbH

Poßmoorweg 2
D-22301 Hamburg
Phone +49 (0) 221-37 75-0
Fax +49 (0) 221-37 75-390
Email: building.de@omya.com
www.omya.de

Member of IVK

Company

Contact
Gabriele Bender
gabriele.bender@omya.com

Range of Products

Raw materials
Additives:
Dispersants for water borne systems
Rheology Modifier (PU-/Acrylic thickeners,
Bentonites, Sepiolites)
Super Plasticizer (Dry Polycarboxylate Ether)

Industrial Minerals:
Calcium Carbonates, Dolomite,
Ultrafine PCC
Kaolin, Baryte
Flame Retardants (ATH)
Lightweight Fillers

Polymers:
Vinyl Acetate (VAC)
Acrylic and Styrene-Acrylic Copolymer
Alkyd and Polyester Resins
Styrene Butadiene Rubber (SBR)
Epoxy Resins, Hardeners and Reactive
Diluents
Hot Melt Resins (PA)

For applications in the field of
Paper/packaging
Bookbinding/graphic design
Wood/furniture industry
transportation-/automotive-/aerospace
industry
Construction industry, including floors,
walls and ceilings
Sealants
Casting-, potting materials
Textile industry
Household, recreation and office

Organik Kimya Netherlands B. V.

Chemieweg 7, Havenummer 4206
NL-3197KC Rotterdam Botlek,
Phone +31 10 295 48 20
Fax +31 10 295 48 29
Email: organik@organikkimya.com
www.organikkimya.com

Member of IVK

Company

Year of formation
1924

Size of workforce
500

Managing partners
Simone Kaslowski
Stefano Kaslowski

Nominal capital
100 %

Ownership structure
100 % familiy owned business

Subsidiaries
Distribution worldwide, production sites in
the Netherlands and Turkey

Sales channels
Direct sales and through distributors
worldwide

Contact partners
Management:
Stefano Kaslowski, General Manager

Application technology and sales:
Oguz Kocak, Sales Manager
Phone: +49 (0) 173-6 52 22 59
Email: o_kocak@organikkimya.com

Range of Products

Types of adhesives
Dispersion adhesives
Pressure-sensitive adhesives

Types of sealants
Acrylic sealants

Raw materials
Polymers

For applications in the field of
Paper/packaging
Bookbinding/graphic design
Wood/furniture industry
Construction industry, including floors,
walls and ceilings
Automotive industry, aviation industry
Textile industry
Adhesive tapes, labels

Sealants • Adhesives

Hermann Otto GmbH
Krankenhausstraße 14
D-83413 Fridolfing
Phone +49 (0) 86 84-9 08-0
Fax +49 (0) 86 84-9 08-5 39
Email: info@otto-chemie.com
www.otto-chemie.com

Member of IVK

Company

Management board
Johann Hafner
Matthias Nath

Size of workforce
400 employees

Contact
Technical Departement
Phone +49 (0) 8 68 49 08-4 60
Email: industry@otto-chemie.de

Sales manager industrial applications:
Marc Wüst
Phone +49 (0) 8 68 49 08-5 21
Email: marc.wuest@otto-chemie.de

Range of Products

Types of sealants and adhesives
1-comp. and 2-comp. silicones
1-comp. and 2-comp. polyurethanes
1-comp. and 2-comp. hybrids
acrylates

For applications in the field of
Renewable energies
Household appliance industry
Lighting and electronics
Textile industry
HVAC (heating, ventilation + air conditioning systems)
Sandwich and composite panels

Panacol-Elosol GmbH
Member of Hönle Group
Daimlerstraße 8
D-61449 Steinbach/Taunus
Phone +49 (0) 61 71 62 02-0
Fax +49 (0) 61 71 62 02-5 90
Email: info@panacol.de
www.panacol.com

Member of IVK

Company

Year of formation
1978

Size of workforce
50

Managing Director
Florian Eulenhöfer

Nominal capital
255,645 €

Ownership structure
Panacol-Elosol GmbH is the German subsidiary of Panacol AG in Switzerland. Panacol Group is the subsidiary of Dr. Hönle AG

Sales channels
Sales team for Germany
International distribution network with worldwide sales partners

Contact partners
Sales Director and Application Engineering: Dr. Detlef Heindl

Further information
Panacol is a leading international manufacturer of industrial adhesives as well as medical grade adhesives. In addition to Eleco-EFD, an affiliated company in France, Panacol provides an international network of distributors, which ensure a personal advisory service around the world. Since January 2008 Panacol has been a member of the Hönle Group and is benefiting from the numerous synergies in the field of industrial UV technology.

Range of Products

Types of adhesives
Reactive adhesives
Anaerobic adhesives
Cyanoacrylates
LED/UVA/visible light curing epoxy and acrylate adhesives
High temperature adhesives
Conductive adhesives
Medical grade adhesives
Structural adhesives

Types of sealants
Epoxies
Acrylics

Equipment
UV and LED equipment for adhesive curing from Hönle UV technology

Market Segments
Electronics/Applications on PCBs
Conformal coatings
Smart card/Die attach
Optics and fibre optics/Active alignment
Optoelectronics
Medical device assembly
Display bonding/Liquid gaskets
Loudspeaker assembly
Appliances
Automotive and Aviation Industry

experience. **performance.**

Paramelt B.V.
Costerstraat 18
NL-1704 RJ Heerhugowaard
Phone +31 (0) 72 5 75 06 00
Fax +31 (0) 72 5 75 06 99
Email: info@paramelt.com
www.paramelt.com

Member of VLK

Company

Year of formation
1898

Size of workforce
400

Ownership structure
privately owned

Subsidiaries
Paramelt Veendam B.V.; Paramelt USA Inc.;
Paramelt Specialty Materials (Suzhou) Co., Ltd.

Sales channels
Sales offices in Sweden, Germany,
Netherlands, UK, France and Portugal and
a network of specialised distributors

Contact partners
BU Packaging: Neill Dutton
BU Construction & Assembly: Wim van Praag

About Paramelt
Paramelt (established in 1898) has grown over
the years to become the leading global specialist in wax based materials and adhesives.
Today Paramelt operates from 7 production
locations around the globe in The Netherlands, USA and China. The company functions
through a series of global business units providing a structured approach to the key market
sectors in which we operate like e.g. packaging
and construction & assembly. For these markets we offer a comprehensive range of waxes,
water based adhesives, hot melt (pressure
sensitive) adhesives, water-based functional
coatings, solvent-based and PU adhesives.
Serviced by both regional sales offices, as well
as a comprehensive network of distribution

Range of Products

Types of adhesives
Hot melt adhesives
Reactive adhesives
Solvent-based adhesives
Dispersion adhesives
Vegetable adhesives, casein, dextrin and
starch adhesives
Pressure-sensitive adhesives
Heat seal coatings

Types of sealants
Other

For applications in the field of
Paper/(Flexible) Packaging/Labelling
Construction industry/General Assembly

partners, our customers can be assured of the
highest levels of local service and support.
Paramelt possess extensive experience in the
design and development of adhesives and
functional coatings to meet critical machine
and application requirements. The company
has built significant knowledge of performance
aspects needed to make our products effective
at all stages of the supply chain. Our products
are backed up by regional laboratories providing
comprehensive application and analytical testing facilities to ensure selection of the most
appropriate adhesive for your application.
Built on a tradition of partnership and trust;
underpinned by detailed knowledge gained
over more than 100 years of operation, Paramelt can offer real benefits to your operation.

PCI Augsburg GmbH

Piccardstraße 11
D-86159 Augsburg
Phone +49 (0) 8 21 59 01-0
Fax +49 (0) 8 21 59 01-3 72
Email: pci-info@basf.com
www.pci-augsburg.com

Member of IVK, FCIO

Company

Year of formation
1950

Size of workforce
800

Ownership structure
PCI Augsburg GmbH is part of BASF –
The Chemical Company

Subsidiaries
please refer to website for details

Sales channels
indirect / through distributors

Contact partners
Management:
please refer to website

Application technology and sales:
please refer to website

Further information
please refer to website

Range of Products

Types of adhesives
Reactive adhesives
Dispersion adhesives

Types of sealants
Acrylic sealants
PUR sealants
Silicone sealants

Equipment, plant and components
or conveying, mixing, metering and
for adhesive application

For applications in the field of
Construction industry, including floors,
walls and ceilings

PLANATOL®
smart gluing

Planatol GmbH
Fabrikstraße 30 – 32
D-83101 Rohrdorf
Phone +49 (0) 80 31-7 20-0
Fax +49 (0) 80 31-7 20-1 80
Email: info@planatol.de
www.planatol.de

Niederlassung Herford:
Hohe Warth 15 – 21
D-32052 Herford
Tel.: +49 (0) 52 21-77 01-0
Fax: +49 (0) 52 21-715 46
Email: info@planatol.de
www.planatol.de

Member of IVK

Company

Year of formation
1932

Size of workforce
130

Ownership structure
Blue Cap AG

Sales channels
Direct sale to industrial customers
Own sales staff and sales representatives Retailers
World-wide sales and service network
Own sales force, distributors

Contact partners
Managing director:
Johann Mühlhauser
Phone +49 (0) 80 31-720-0

Further information
Planatol GmbH is a member of Blue Cap AG,
which bundles a wide range of competence
in print and adhesive technology: High-tech
fold-gluing systems for rotary printing usable for
dispersions and hotmelt (Planatol System), post
print and glue logistic machines (Gämmerler),
Copy Binder (Planax), high quality coated self-
adhesive and digital print media (Neschen) and
other consumer and investment goods.

Range of Products

Types of adhesives
Hot-melt adhesives
Dispersion adhesives
Dextrin and
starch adhesives
Pressure-sensitive adhesives
(PSA) both dispersion and hotmelt
UF-resins and accelerators
EVA and PO
Metallocene
APAO Hotmelts
Reactive PUR hotmelts and PU dispersionen
Planamelt

For applications in the field of
Graphic arts
– Bookbinding
– Print finishing
– Foldgluing
– Specialities

Packaging
– Folding box
– End of line
– Bags and sacks
– Labels
– Specialities

Wood adhesives
– Kitchen and furniture
– Wood composites
– Construction

Other
– Walls and ceilings
– Automotive industry, aviation industry
– Adhesive tapes
– Hygiene
– Sandwich elements
– Specialities

POLY-CHEM GmbH
ChemiePark Bitterfeld-Wolfen
OT Greppin, Farbenstraße, Areal B
D-06803 Bitterfeld-Wolfen
Phone +49 (0) 3493 75400
Fax +49 (0) 3493 75404
Email: info@poly-chem.de
www.poly-chem.de

Member of IVK

Company

Year of formation
2008

Size of workforce
50

Sales channels
Direct sales to industrial partners

Contact partners
Management:
Dr. Jörg Dietrich

Application technology and sales:
Dr. Sandra Starke
Dr. Andreas Berndt

Further information
Toll production, Custom Synthesis

Range of Products

We offer to the coating industry, to the chemical industry and as well to the segments paints and varnishes efficient and innovative solutions.

The present business activities of POLY-CHEM AG are
- Solvent based and solvent-free pressure sensitive acrylic adhesives and acrylic polymers in general
- Synthetic rubber pressure sensitive adhesive solution
- Production of specialty chemicals
- Contract formulations
- Services such as drum filling and tank leasing
- Trading (export/import)

Raw materials
Additives: Crosslinker, softening agents
Polymers: Polyacrylates

For applications in the field of
- Paper/packaging
- Construction industry, including floors, walls and ceilings
- Textile industry
- Adhesive tapes, labels

Polymere Technologien

Polytec PT GmbH
Polymere Technologien
Polytec-Platz 1 – 7
D-76337 Waldbronn
Phone +49 (0) 72 43 - 604 - 40 00
Fax +49 (0) 72 43 - 604 - 42 00
Email: info@polytec-pt.de
www.polytec-pt.de

Member of IVK

Company

Year of formation
2005 (1967)

Size of workforce
20

Contact partners
Management:
info@polytec-pt.de

Application technology and sales:
info@polytec-pt.de

Further information
Polytec PT GmbH develops, manufactures and distributes special adhesives for applications in electronics, electrical engineering and automotive electronics as well as the solar industry and the manufacture of smart cards. In addition to an extensive portfolio of electrically and thermally conductive adhesives, transparent and UV-hardening products, Polytec PT develops tailored formulations for the most demanding of adhesive applications.

Range of Products

Range of Products
Electrically conductive adhesives
Thermally conductive adhesives
Adhesives for optical assemblies / fiber optics
USP Class VI adhesives for medical devices
High temperature adhesives
Potting compounds
UV-curable adhesives
Surface pretreatment devices

For applications in the field of
Electronics
Mechanical engineering and equipment construction
Automotive industry, aviation Industry

PRHO-CHEM GmbH

Dohlenstraße 8
D-83101 Rohrdorf-Thansau
Phone +49 (0) 80 31-3 54 92-0
Fax +49 (0) 80 31-3 54 92-29
Email: info@prho-chem.de
www.prho-chem.de

Member of IVK

Company

Year of formation
1994

Ownership structure
private property

Contact partners
Management:
Otto Kleinhanß

Application technology and sales:
Otto Kleinhanß

Range of Products

Types of adhesives
Hot melt adhesives
Reactive adhesives
Dispersion adhesives
Glutine glue
Pressure-sensitive adhesives

Types of sealants
PUR sealants
Silicone sealants
MS/SMP sealants

For applications in the field of
Paper/packaging
Graphic design
Automotive industry, aviation industry
Hygiene
Adhesive tapes, labels
Industrial applications

RAMPF Polymer Solutions GmbH & Co. KG

Robert-Boschstraße-Straße 8 – 10
D-72661 Grafenberg
Phone +40 (0) 71 23-9342-0
Fax +49 (0) 71 23-93 42-2444
Email: polymer.solutions@rampf-gruppe.de
www.rampf-gruppe.de

Member of IVK

Company

Year of formation
1980

Size of workforce
90

Managing partners
Dr. Klaus Schamel

Ownership structure
Family owned

Contact partners
Management:
Dr. Klaus Schamel

Application technology and sales:
Dr. Wolfgang Hodek

Further information
www.rampf-gruppe.de

Range of Products

Types of adhesives
Hot melt adhesives
Reactive adhesives

Types of sealants
PUR sealants
Silicone sealants
MS/SMP sealants

For applications in the field of
Wood/furniture industry
Construction industry, including floors,
walls and ceilings
Electronics
Mechanical engineering and equipment
construction
Automotive industry
Transportation industry
Filter industry
Household appliances

Ramsauer GmbH & Co. KG
Sarstein 17
A-4822 Bad Goisern
Phone +43 (0) 61 35-82 05
Fax +43 (0) 61 35-83 23
Email: office@ramsauer.at
www.ramsauer.at

Member of IVK

Company

Year of formation
1875

Head of Business Administration
Klaus Forstinger, Head of Business
Administration

Ownership structure
private

Contact partners
Management:
Andreas Kain, Sales Manager

Further information
Company history:
When Ferdinand Ramsauer purchased a small
chalk quarry near Bad Goisern in 1875, he
already had all the skills typical of successful
people up to our days. He had his mind set
on innovation and was fully focussed on achieving his goals. Within less than 20 years, he
increased the chalk output of his quarry about
100 times and made "Ischler Bergkreide"
("Mountain chalk from Bad Ischl") a well-known
brand name. Ferdinand Ramsauer and his son
Josef - whose name our company bears today -
were genuine marketing pioneers.
Right from the beginning, mountain chalk was
used primarily for the production of putty for
glazing. Initially, the raw material was sold
to putty manufacturers. In 1950, Ramsauer
began manufacturing putty on its own. The
evolution from mining operations to sealant
manufacturer was complete.
With the introduction of thermally insulating
windows, new plastic and elastic sealants
were needed. Ramsauer developed the first
such modified putties as early as in the fif-
ties. Between 1960 and 1976, sealants were
launched under the very well known names of

Range of Products

Types of adhesives
Reactive adhesives
(1 and 2 component solutions)
Solvent-based adhesives
Dispersion adhesives

Types of sealants
Acrylic sealants
Butyl sealants
PUR sealants
Silicone sealants
MS/SMP sealants

For applications in the field of
Wood/furniture industry
Construction industry, including floors,
walls and ceilings
Mechanical engineering and equipment
construction
Automotive industry, aviation industry
Hygiene and clean room application
Household, recreation and office

E9 M, HB, R68, HV72, and HV76. Simultane-
ously, Ramsauer started developing the first
water-soluble products, known as acrylates. In
1972, Ramsauer started manufacturing sealants on a silicone basis. In 1976, production of
PU foam was initiated. A patent for 2-compo-
nent systems was registered in 1998.
Currently, Ramsauer manufactures top-quality
sealants of all systems and modifications, including a new silicone-free sealant on a hybrid
basis and a 2-component silicone system.
When looking back at 135 years of company
history, there have been many changes, fundamental changes. However what has remained
is the broad perspective and focus on innovation that lives on in the present generation in
our company.

Renia Gesellschaft mbH
Ostmerheimer Straße 516
D-51109 Köln
Phone +49 (0) 2 21-63 07 99-0
Fax +49 (0) 2 21-63 07 99-50
Email: info@renia.com
www.renia.com

Member of IVK

Company

Year of formation
1930

Ownership structure
GmbH, family owned

Sales channels
exclusive partners in more than 50
countries

Contact partners
Management:
Heinz Buchholz

Application technology:
Dr. Julian Grimme

Exportmanager:
Dr. Rainer Buchholz

Range of Products

Types of adhesives
Solvent-based adhesives
Dispersion adhesives

For applications in the field of
Household, recreation and office
Shoe Industry
Health

Rhenocoll-Werk e. K.

Erlenhöhe 20
D-66871 Konken
Phone +49 (0) 63 84-99 38-0
Fax +49 (0) 63 84-99 38-1 12
Email: info@rhenocoll.de
www.rhenocoll.de

Member of IVK

Company

Year of formation
1948

Size of workforce
120

Ownership structure
Joint partnership, Fam. Holding

Subsidiaries
Polska, Czechoslovakia, Belarus, Georgia,
Russia, China, India

Sales channels
Dealer based worldwide

Contact partners
Management:
Werner Zimmermann

Range of Products

Types of adhesives
Hot melt adhesives
Reactive adhesives
Dispersion adhesives
Pressure-sensitive adhesives

Types of sealants
Acrylic sealants
PUR sealants

For applications in the field of
Paper/packaging
Wood/furniture industry
Construction industry, including floors,
walls and ceilings
Household, recreation and office

RUDERER KLEBETECHNIK GMBH
Harthauser Straße 2
D-85604 Zorneding (Munich)
Phone +49 (0) 8106 2421 -0
Fax +49 (0) 8106 2421 -19
Email: info@ruderer.de
www.ruderer.de

Member of IVK

Company

Year of formation
1987

Size of workforce
> 30

Ownership structure
Family company

Sales channels
direct and distribution

Range of Products

Types of adhesives
Hot melt adhesives
Reactive adhesives
Solvent-based adhesives
Dispersion adhesives
Pressure-sensitive adhesives

Types of sealants
Acrylic sealants
PUR sealants
Silicone sealants
MS/SMP sealants

Equipment, plant and components
for surface pretreatment
for adhesive curing
adhesive curing and drying
measuring and testing

For applications in the field of
Wood/furniture industry
Electronics
Mechanical engineering and equipment
construction
Automotive industry, aviation industry
Textile industry
Household, recreation and office

RÜTGERS
Germany GmbH

Varziner Straße 49
D-47138 Duisburg
Phone +49 (0) 2 03-42 96-02
Fax +49 (0) 2 03-42 96-7 62
Email: resins@raincarbon.com
www.novares.de

Member of IVK

Company

Year of formation
1849

Sales channels
own sales force
global agent and distribution network

Contact partners
Management:
Uwe Holland, Managing Director

Sales Manager
Mai Doan

Range of Products

Raw materials
Hydrocarbon resins:
aromatic
aliphatically mod.
Indene-coumarone
phenolically modified
pure monomer based

Modifiers:
high boiling solvents

SABA Dinxperlo BV

Industriestraat 3
NL-7091 DC Dinxperlo
Phone +31 315 65 89 99
Fax +31 315 65 32 07
Email: info@saba-adhesives.com
www.saba-adhesives.com

Member of VLK

Company

Year of formation
1933

Size of workforce
170

Managing director
W. de Zwart

Ownership structure
R. J. Baruch, W. F. K. Otten

Subsidiaries
SABA Polska SP. z o. o.
SABA North America LLC
SABA Pacific
SABA China
SABA Bocholt GmbH

Contact:
info@saba-adhesives.com

Further information
Whether you are active in construction or industry, the adhesives and sealants that you use have to meet stringent requirements. SABA is a producer of high-quality and technically progressive adhesives and sealants. With our knowledge of adhesion and sealing, we are happy to help you optimise your processes, so that you produce better end-products or projects at lower total costs and in a safer working environment. In this way, we work together to strengthen your competitive position.

Read more about SABA:
www.saba-adhesives.com

Range of Products

Types of adhesives
Water-based adhesives
Hot melt adhesives
Solvent-based adhesives
Reactive adhesives
Pressure-sensitive adhesives

Types of sealants
Polysulfide sealants
PUR sealants
Silicone sealants
MS / SMP sealants

Equipment, plant and components
for conveying, mixing, metering and
for adhesive application
for surface pretreatment

For applications in the field of
furniture production
mattress production
foam converting
automotive
pvc bondings
building & construction
marine
transport
civil & environmental engineering

Schill+Seilacher „Struktol" GmbH
Moorfleeter Straße 28
D-22113 Hamburg
Phone +49 (0) 40-733-62-0
Fax +49 (0) 40-733-62-297
Email: polydis@struktol.de
www.struktol.de

Member of IVK

Company

Year of formation
1877

Size of workforce
250 employees in Hamburg

Ownership structure
privately owned

Subsidiaries
Schill+Seilacher, Böblingen (Germany)
Schill+Seilacher Chemie GmbH,
Pirna (Germany)
Struktol Company of America, Ohio (USA)
Struktol Co. Company of America, Georgia (USA)
SNS Nano Fiber Technology LLC., Ohio (USA)

Sales channels
Germany: Direct
International: Distributeurs and Agencies

Contact partners
Meike Bénet
(Head of Sales Epoxy Products)
Phone: +49 (0) 40-733-62-241
Email: mbenet@struktol.de

Marcel Volstorf
(Application Technology & Project Manager
Epoxy Products)
Phone: +49 (0) 40-733-62-256
Email: mvolstorf@struktol.de

Sven Wiemer
(Senior Manager Epoxy Products)
Phone: +49 (0) 40-733-62-125
Email: swiemer@struktol.de

Further information
Manufacturing and development of tailor-made
solutions of exclusive epoxy prepolymers in
cooperation with our customers.

Range of Products

Raw materials
Epoxy Prepolymers
Struktol® Polydis®
Struktol® Polycavit®
Struktol® Polyvertec®
Struktol® Polyphlox®

The product ranges Struktol® Polydis®,
Polycavit® and Polyvertec® are adducts of either
rubber or elastomer modified epoxy resins to
improve the mechanical properties, such as Impact Resistance, T-Peel and Lap Shear next to a
general improvement of the adhesion behavior.

The Struktol® Polyphlox® range consists of
adducts of organo phosphorous modified epoxy
resins to equip epoxy matrices with flame
retardancy.

Applications
Epoxy based
(Structural) Adhesives
Castings
Prepregs
Composites
Fiber-Reinforced Plastics

Schlüter-Systems KG

Schmölestraße 7
D-58640 Iserlohn/Germany
Phone +49 (0) 23 71-971-0
Fax +49 (0) 23 71-971-11
Email: info@schlueter.de
www.schlueter-systems.com

Member of IVK

Company

Year of formation
1966

Size of workforce
more than 1,000 employees worldwide

Managing partners
Schlüter-Systems KG has system alliances with leading construction chemistry and ceramics companies.

Our constructions chemistry partners are: ARDEX GmbH, PCI Augsburg GmbH, Sopro Bauchemie GmbH, MAPEI GmbH, Kiesel Bauchemie GmbH u. Co. KG and SCHÖNOX GmbH, RAK Ceramics GmbH and V&B Fliesen GmbH.

Subsidiaries
Apart from its head office in Iserlohn, Germany, the company has seven subsidiaries in Europe and North America as well as several service bureaus and distribution partners around the world. In total the company employs more than 1,000 people.

Sales channels
B2B

Contact partners
Management:
Günter Broeks (Sales Director)

Application technology:
Rainer Reichelt
(Head of International Technical Network)

Further information
From the Schlüter-SCHIENE to complete systems – with innovative ideas and high-quality products Schlüter-Systems KG is the market leader for many tile related products.

Range of Products

Types of adhesives
Reactive adhesives
Dispersion adhesives

Types of sealants
Other

Raw materials
Polymers

Equipment, plant and components
for surface pretreatment
for adhesive curing
adhesive curing and drying

For applications in the field of
Construction industry, including floors, walls and ceilings
Adhesive tapes, labels
Hygiene

Schlüter-Systems KG with its headquarter in Iserlohn, Germany, offers complete systems with a portfolio of more than 10,000 products and product types and has distribution partners around the world. Among other things, the family-run company offers drainage systems for balconies and terraces, underfloor heating systems, barrier-free showers, construction panels, uncoupling and insulation solutions and a multitude of profiles. Innovative illuminated profile technology and wall heating systems enlarge Schlüter-Systems KG product portfolio.

With more than 1,000 employees in Europe and the USA Schlüter-Systems KG sets standards in creating solutions for tile setters worldwide.

A SIKA BRAND

Sika Deutschland GmbH
Marke Schönox
Alfred-Nobel-Straße 6
D-48720 Rosendahl
Phone +49 (0) 25 47-9 10-0
Fax +49 (0) 25 47-9 10-1 01
Email: info@schoenox.de
www.schoenox.com

Member of IVK, VLK

Company

Year of formation
1891

Size of workforce
340

Ownership structure
Sika Holding GmbH

Sales channels
Wholesalers for tiling, floor coverings and general building materials

Application technology and sales
Quality products for installing all types of tiles, wall and floor coverings for professional

Application technology and sales
Sika Deutschland GmbH is a subsidiary of Sika AG and part of the Sika Area Germany. The company develops and manufactures quality products for installing all types of tiles and floor coverings for professional craftsmen. The assortment includes tile adhesives, floor adhesives, grouts, levelling compounds, screeds, primers and waterproof membranes. SCHÖNOX products are distributed in more than 20 European countries and North America. About 340 employees work at the Rosendahl site in Germany.

Range of Products

Types of adhesives
Dispersion adhesives

Types of sealants
Silicone sealants
MS/SMP sealants

For applications in the field of
Construction industry, including floors, walls and ceilings

Schomburg GmbH & Co. KG

Aquafinstraße 2 – 8
D-32760 Detmold
Phone +49 (0) 52 31-9 53-00
Fax +49 (0) 52 31-9 53-1 23
Email: info@schomburg.de
www.schomburg.de

Member of IVK

Company

Year of formation
1937

Size of workforce
220 (Germany), 580 worldwide

Managing partners
Albert Schomburg
Ralph Schomburg
Alexander Weber

Nominal capital
3,619 Mio. €

Ownership structure
Family and Management owned

Subsidiaries
31 worlwide:
Poland, Czech Republic, USA, India, Turkey,
Luxemburg, Switzerland, Russia, Netherlands, Slovakia, Italy, etc.

Sales channels
Distribution partners

Contact partners
Management:
Ralph Schomburg
Alexander Weber

Application technology and sales:
Holger Sass
Michael Hölscher

Range of Products

Types of adhesives
Reactive adhesives
Dispersion adhesives
Cement based adhesives

Types of sealants
Acrylic sealants
Polysulfide sealants
PUR sealants
Silicone sealants
Other

Equipment, plant and components
for conveying, mixing, metering and
for adhesive application

For applications in the field of
Construction industry, including floors,
walls and ceilings
Mechanical engineering and equipment
construction

SIEMA Industrieklebstoffe GmbH
Eichelsbacher Straße 6
D-66954 Pirmasens-Gersbach
Phone +49 (0) 63 31-9 15 67
Fax +49 (0) 63 31-9 23 89
Email: Verkauf@siema-industrieklebstoffe.de
www.siema-industrieklebstoffe.de

Member of IVK

Company

Year of formation
1975

Size of workforce
15

Nominal capital
52,000

Ownership structure
privat

Sales channels
trade and direct

Contact partners
Management:
Johannes Illik

Application technology and sales:
Johannes Illik
Marco Dimitrijevic

Further information
Flexible company that specializes in custom
solutions and customer inquiries

Range of Products

Types of adhesives
Hot melt adhesives
Reactive adhesives
Solvent-based adhesives
Dispersion adhesives
Vegetable adhesives, dextrin and starch
adhesives
Glutine glue
Pressure-sensitive adhesives

For applications in the field of
Paper/packaging
Bookbinding/graphic design
Wood/furniture industry
Electronics
Mechanical engineering and equipment
construction
Automotive industry, aviation industry
Textile industry
Household, recreation and office

BUILDING TRUST

Sika Automotive GmbH
Reichsbahnstraße 99
D-22525 Hamburg
Phone: +49 (0) 40-5 40 02-0
Fax: +49 (0) 40-5 40 02-5 15
Email: info.automotive@de.sika.com
www.sika-automotive.de

Member of IVK

Company

Year of formation
1928

Size of workforce
260

Subsidiaries
Sister companies in 97 countries

Contact partners
Managing Director:
James Miko

Export:
Kai Paschkowski

Range of Products

Types of adhesives
Hot melt adhesives
Reactive adhesives
Solvent-based adhesives
Dispersion adhesives
Pressure-sensitive adhesives

Types of sealants
PUR sealants
Other

For applications in the field of
Electronics
Automotive industry
Textile industry
Adhesive tapes & labels
Hygiene

**Creating Solutions
for Increased Productivity**
Sika is supplier and development partner to
the automotive industry. Our state-of-the-art
technologies provide solutions for increased
structural performance, added acoustic
comfort and improved production processes. As a specialty company for chemical
products, we concentrate on our core
competencies: **Bonding – Sealing –
Damping – Reinforcing**
As a globally operating company, we are
partner to our customers worldwide. Sika
is represented with its own subsidiaries in
all automobile-producing countries, thus
guaranteeing a professional and fast local
service.

BUILDING TRUST

Sika Deutschland GmbH
Stuttgarter Straße 139
D-72574 Bad Urach
Phone +49 (0) 71 25-940-761
Fax +49 (0) 71 25-940-763
Email: industry@de.sika.com
www.sika.de/industrie

Member of IVK, FKS, VLK

Company

Year of foundation
1910

Size of workforce
17,500 (worldwide Sika AG),
1,500 Sika Deutschland GmbH

Subsidiaries
in more than 90 countries,
see www.sika.com

Sales channels
direct and distribution

Range of Products

Industry and sealants
PUR adhesives
Reactive adhesives
Dispersion adhesives
Solvent-based adhesives
Hot melt adhesives
Pressure-sensitive adhesives
Epoxy adhesives
Acrylic adhesives and sealants
Laminating adhesives
Silicones
SMP adhesives and sealants

Construction
PUR adhesives and sealants
Polysulfide sealants
Acrylic adhesives and sealants
Silicones
SMP adhesives and sealants
Butyl sealants

For applications in the field of
Mechanical engineering and equipment
construction
Construction industry, including floors,
walls and ceilings
Automotive industry, aviation industry
Wood/furniture industry
Electronics
Adhesive tapes, labels
Household, recreation and office
Photovoltaics
Renewable Energies
Marine
Structural glazing
Facades
Building Constructions
Track Construction

BUILDING TRUST

Sika Nederland B.V.
Zonnebaan 56
NL-3542 EG
Phone +31 (0) 30-241 01 20
Fax +31 (0) 30-241 01 20
Email: info@nl.sika.com
www.sika.nl

Member of VLK

Company

Size of workforce
140

Subsidiaries
2

Contact partners
Management:
info@nl.sika.com

Application technology and sales:
info@nl.sika.com

Range of Products

Types of adhesives
Hot melt adhesives
Solvent-based adhesives
Dispersion adhesives

Types of sealants
Acrylic sealants
PUR sealants
Silicone sealants
MS/SMP sealants

Equiment, plant and components
for surface pretreatment

For applications in the field of
Wood/furniture industry
Construction industry, including floors,
walls and ceilings
Electronics
Mechanical engineering and equipment
construction
Automotive industry, aviation industry

Sonderhoff Chemicals GmbH
Richard-Byrd-Straße 26
D-50829 Cologne
Phone +49 221 95685-0
Fax +49 221 95685-599
Email: info@sonderhoff.com
www.sonderhoff.com

Company

Year of formation
1958

Size of workforce
95 employees

More than 260 employees within Sonderhoff Group of Companies worldwide

Ownership structure
Sonderhoff Holding GmbH

Sister companies
Sonderhoff Engineering GmbH,
Dornbirn (Austria)
Sonderhoff Services GmbH,
Cologne (Germany)
Sonderhoff Polymer-Services Austria GmbH,
Hörbranz (Austria)
Sonderhoff Italia SRL, Oggiono (Italy)
Sonderhoff USA Corporation, Elgin (USA)
Sonderhoff (Suzhou) Sealing Systems
Co.Ltd, Suzhou (China)

Sales channels
worldwide

Contact partners
Daniel Koscielny (sales director)

Further information
Sonderhoff, the specialist for the Formed In-place sealing technology, offers everything from one hand on: more than 1,000 recipes for foam sealing, potting and gluing material, plant engineering adapted to customer specific manufacturing processes, as well as multiple patented knowledge and skills experiences with a variety of realized applications from more than 55 years.

Range of Products

Types of adhesives
2-Component reaction adhesives based on PU, available in different degrees of hardness and viscosities (liquid to stable). Good adhesion on many substrates. Good temperature and climate change resistance, compliant to automotive requirements. Also used in others technical areas and sectors.

Types of sealants
2C PU, 2C Silicone, 1C PVC

Systems/Procedures/Accessories/Services
Order systems (1C systems, 2C/multi-component systems, robots), automation for mixing and dosing technology, for adhesive applications, services, contract manufacturing

For applications in the fields
Automotive, packaging, electronics, aviation-, filter- and lighting industry, machine and device construction, household (white goods), photovoltaics, solar thermal energy, air condition, switch board enclosures

Sopro Bauchemie GmbH
Biebricher Straße 74
D-65203 Wiesbaden
Phone +49 611-1707-239
Fax +49 611-1707-240
Email: international@sopro.com
www.sopro.com

Member of IVK, FCIO

Company

Year of formation
1985 as Dyckerhoff Sopro GmbH, in the year 2002 change in Sopro Bauchemie GmbH

Size of workforce
295 employees

Managing partners
Michael Hecker, Andreas Wilbrand

Subsidiaries
Germany, Hungary, Switzerland, Netherland, Poland, Austria

Sales channels
through specialised distributors

Contact partners
Management:
International Business
Phone +49 611-1707-239
Fax +49 611-1707-240
Email: international@sopro.com

Further information
Sopro offers a comprehensive range of tile-fixing and building chemicals products. Our clear-cut brand strategy has established us as a leading specialist in this sector. Sopro's wide-ranging product portfolio, featuring a consistently high proportion of new products, addresses the full gamut of tiling and building chemicals applications while quaranteeing a product quality that complies, in all respects, with the strict standards set by professional applicators. The builders' merchants sector represents the only acceptable marketing channel for the Sopro brand, Merchants are able to provide both professional tradesmen and the ambitious private client with the standard of counselling appropriate to our high-grade, technologically advanced products.

Range of Products

Types of adhesives
Cementitious adhesives
Reactive adhesives
Dispersion adhesives

Types of sealants
Acrylic sealants
PUR sealants
Silicone sealants

For applications in the field of
Construction industry, including floors, walls and ceilings

Our product range is divided into thress sectors:
Tile fixing products: Tile Adhesives; Tile grouts; Surface fillers; Primers and bonding agents; Waterproofings; Accessories (impact sound insulation and seperating layer systems); Cleaning, impregnation and maintenance; Tiling tools

Building chemical products: Bitumen products; Screeds, binders and construction resins; Mortar and screed additives; Bedding and multi-purpose mortars; Underground construction, gully-repair and shrinkage compounded grouts; Surface fillers; Concrete repairs; Primers, bonding agents, cement paints and silicia sands

Gardening and landscaping products: Drainage and bedding mortars; Paving grouts; Waterproofings; Surface gradient fillers and rapid set mortars; Cleaning, impregnation and maintenance; Landscaping product systems

Stauf Klebstoffwerk GmbH

Oberhausener Str. 1
D-57234 Wilnsdorf
Phone +49 (0) 27 39-3 01-0
Fax +49 (0) 27 39-3 01-2 00
Email: info@stauf.de
www.stauf.de

Member of IVK, FCIO

Company

Year of formation
1828

Size of workforce
72

Ownership stucture
100 % Family Stauf

Managerial head
Volker Stauf, Wolfgang Stauf

Sales channels
Worldwide distribution, own field service and distributing warehouses in Germany and other countries for:
- the wood flooring wholesale
- the floor covering wholesale
- the construction material wholesale
- handcraft enterprises
- contractors
- architects

Products
- adhesive systems for floor covering and wood flooring
- mounting repair adhesives
- artificial turf adhesives
- primers
- levelling compounds
- underlayments
- surface treatment products
- accessories

Contact partners
Management:
Volker Stauf, Phone +49 (0) 27 39-30 10,
info@stauf.de

Product Technology:
Dr. Frank Gahlmann, Phone +49 (0) 27 39-3 01-1 65, gahlmann@stauf.de

Range of Products

Types of adhesives
Reactive adhesives
Solvent-based adhesives
Dispersion adhesives

Types of surface treatment products
waterborne finish systems
conventional finish systems
Oils
Care + cleaning systems

For applications in the field of
Construction industry, including floors, walls and ceilings
- Parquet and wood flooring
- End grain wood blocks
- Textile and elastic floor coverings
- Artificial turf
- Mounting repair
- Industrial application

Further information
www.stauf.de

STAUF is a leading system supplier for flooring technology. For the safe and durable bonding of wood flooring and floor coverings we research, develop and produce innovative adhesive systems on a high-grade raw material basis. Next to the latest adhesive systems STAUF offers professionel surface treatment products for wooden surfaces as well as the whole bandwidth of products for sub floor preparation and accessories.

STAUF has remained a family-owned company even after more than 185 years.

Long time experience as well as continuous advancement in a state of the art production and research environment ensure the constant top-level product quality and set the standards for the customers of the wood flooring and floor covering branch.

Stockmeier Urethanes GmbH & Co. KG
Im Hengstfeld 15
D-32657 Lemgo
Phone +49 (0) 52 61-66 0 68-0
Fax +49 (0) 52 61-66 0 68-29
Email: urethanes.ger@stockmeier.com
www.stockmeier-urethanes.com

Member of IVK

Company

Year of formation
1991

Size of workforce
approx. 150

Ownership structure
Member of Stockmeier Group

Subsidiaries
Stockmeier Urethanes USA Inc., Clarksburg/USA, Stockmeier Urethanes France S.A.S.,Cernay, Stockmeier Urethanes UK Ltd., Sowerby Bridge/UK

Sales channels
direct and distribution

Contact partners
Management:
Stefan Baumann

Application technology and sales:
Frank Steegmanns

Further information
Stockmeier Urethanes is a leading international manufacturer of polyurethane systems. We are a specialized subsidiary of the family-owned Stockmeier Group. With four production plants including R & D in Europe and the USA we are developing and producing polyurethane systems for sports and

Range of Products

Types of adhesives
Reactive adhesives

Types of sealants
PUR sealants

For applications in the field of
Wood/furniture industry
Construction industry, including floors, walls and ceilings
Electronics
Mechanical engineering and equipment construction
Automotive industry, aviation industry

leisure flooring, ACE (Adhesives, Coatings & Elastomers) and electrical encapsulation since 1991. In our business unit Adhesives we are producing systems for manufactures in the markets Batteries, Filters, Sandwichpanel, Transportation, Caravan or customized products for other industrial applications. Our top brands are Stobielast, Stobicast, Stobicoll and Stobicoat. More information: www.stockmeier-urethanes.com

Synthomer Deutschland GmbH
Werrastraße 10
D-45768 Marl
Email: info.europe@synthomer.com
www.synthomer.com

Company

Company

Synthomer is one of the world's leading suppliers of emulsion and speciality polymers supporting leadership positions in many market segments including coatings, construction, technical textiles, adhesives, paper and synthetic latex gloves.

Our strategy is based first and foremost on customer service, provided worldwide through a strong network of local sales and technical service, supported by regional application development and production in our key markets. The company has its headquarters in London, UK and provides customer focused services from operational centres in Harlow (UK), Marl (Germany), Kuala Lumpur (Malaysia) and Shanghai (China). We supply our customers from more than 20 manufacturing sites across the world.

Contact

Dr. Katja Greiner
European Technical Sales Manager Adhesives
SBU Functional Solutions
Phone: +49 (0) 2365-49 9816
Mobile: +49 (0) 171-813 7422
Email: katja.greiner@synthomer.com

Ying Ho Lee
Marketing Manager SBU Functional Solutions
Phone: +49 (0) 2365-49 19 875
Mobile: +49 (0) 151-2032 0583
Email: yingho.lee@synthomer.com

Range of Products

Raw materials

Dispersions for adhesive tapes, labels, protective foils, wood adhesives, packaging adhesives, sealants

Pure Acrylics *Plextol*®

Styrene Acrylics *Revacryl*®

Vinylacetate
Homo- and Copolymers *Emultex*®

Additives

Thickener *Rohagit*®

Synthopol Chemie
Alter Postweg 35
D-21614 Buxtehude
Phone +49 (0) 41 61-70 71 962
Fax +49 (0) 41 61-8 01 30
Email: bprueter@synthopol.com
www.synthopol.com

Member of IVK

Company

Year of formation
1957

Size of workforce
190

Ownership structure
Family company

Sales channels
Germany and Europe

Contact partners
Management:
Dr. Rüdiger Spohnholz
Phone +49 (0) 41 61-70 71 160
Dr. Stephan Reck
Phone: +49 (0) 41 61-70 71 130

Sales:
Hubert Starzonek (Commercial Manager)
Phone +49 (0) 41 61-70 71 770

Application technology:
Dipl. Ing. Rainer Jack
Phone +49 (0) 41 61-70 71 171

Further information
Birgit Prüter
Phone +49 (0) 41 61-70 71 962

Range of Products

Raw materials
acrylic emulsions
polyurethane dispersions
saturated polyesters
solvent based and waterbased acrylics

For applications in the field of
construction adhesives
automobile adhesives
textile adhesives
pressure sensitive adhesives

TER Chemicals Distribution Group

Börsenbrücke 2
D-20457 Hamburg
Phone +49 (0) 40-30 05 01-0
Fax +49 (0) 40-33 50 50
Email: info@terchemicals.com
www.terchemicals.com

Member of IVK

Company

Year of formation
1908

Size of workforce
340

Managing partners
Christian Westphal

Equity
40 Mio. €

Subsidiaries
30

Sales channels
Trading, Distribution, Salesforce

Contact partners
Management:
Andreas Früh, CEO

Application technology and sales:
Jens Vinke

Further information
www.terchemicals.com

Range of Products

Types of adhesives
Hot melt adhesives
Reactive adhesives
Solvent-based adhesives
Dispersion adhesives
Pressure-sensitive adhesives

Types of sealants
Butyl sealants
PUR sealants
PIB

Raw materials
Additives
Fillers
Resins
Solvents
Polymers

For applications in the field of
Paper/packaging
Bookbinding/graphic design
Wood/furniture industry
Construction industry, including floors,
walls and ceilings
Electronics
Automotive industry, aviation industry
Textile industry
Adhesive tapes, labels
Hygiene

tesa SE

Quickbornstraße 24
D-20253 Hamburg
Phone +49 (0) 40-49 09-1 01
Fax +49 (0) 40-49 09-60 60
www.tesa.de
www.tesa.com

Member of IVK

Company

Year of formation
tesa AG 2001,
tesa SE since march, 30th 2009

Size of workforce
4,100

Ownership structure
100 % subsidiary of Beiersdorf AG, Hamburg

Subsidiaries
54

Sales channels
Industry (e. g. automotive, electronics,
print & paper, solar), Food and DIY

Contact
www.tesa.com

Further information
tesa SE is one of the world's leading
manufacturers of technical adhesive tapes
and self-adhesive system solutions. The
company's focus is on innovative problem-
solving for industrial customers and consumers. The spectrum of applications in the
industrial field ranges from special adhesive
tapes for the printing and paper industries
via cable loom tapes for cars and high-
performance products for fixing electronic
components in mobile phones and digital
cameras to forgery-proof laser labels.
tesa became world famous for branded
products for end consumers – like tesa
Powerstrips® or tesafilm®, one of the few
brand names to be listed in the Duden

Range of Products

Types of adhesives
Hot melt adhesives tapes
Reactive adhesives tapes
Pressure-sensitive adhesives tapes

For applications in the field of
Paper/packaging
Bookbinding/graphic design
Wood/furniture industry
Construction industry, including floors,
walls and ceilings
Electronics
Mechanical engineering and equipment
construction
Automotive industry, aviation industry
Adhesive tapes, labels
Household, recreation and office

dictionary. 300 professional aids are available
to consumers at DIY and discount stores
for creating their personal environment at
work, at home or in the garden.
tesa SE is a member of the Beiersdorf
Group and has been an independent public
company since 2001. With a total staff of
4,100 in 54 affiliates, tesa is active in more
than 100 countries worldwide.

Find out more about tesa SE at www.tesa.
com

tremco illbruck GmbH & Co. KG
Von-der-Wettern-Straße 27
51149 Köln
Phone: +49 (0) 22 03-57 55-0
Fax +49 (0) 22 03-57 55-90
Email: industry@tremco-illbruck.com
www.tremco-illbruck.com

Member of VLK

Company

Year of formation
tremco 1928

Size of workforce
> 1,000

Sales channels
- Distributors (Construction business)
- Direct sales to Key Account Customers
  in the manufacturing industry and to EIFS
  and Window & Facade System Providers

Contact partners
Business Unit Industrial Solutions:
Andres Klapper

Further information
tremco illbruck is one of Europe's leading
manufacturers of highperformance sealing
and bonding products. Sealing tapes,
membranes, adhesives and sealants with
particular properties form the core of our
extensive range.
The Business Unit Industrial Solutions
specialises in finding individual solutions
for specific requirements from industrial
customers and in optimising their
production processes.

www.tremco-illbruck.com

Range of Products

Types of sealants and adhesives
1-part and 2-part silicones
1-part and 2-part polyurethane
1-part Hybrids
Butyl
Acrylic
Hotmelt

For applications in the field of
Automotive Aftermarket and Transportation
Household appliances
Electronics
Ventilation and air conditioning
Photovoltaic
Window and Facade Systems
Insulating and Structural Glazing
Exterior Insulation and Finish Systems
Construction, incl. window, facade and
interior construction

TSRC (Lux.) Corporation S.a.r.l.

34 – 36 Avenue de la Liberté
L-1930, Luxembourg
Phone + 352 - 26 29 72 - 1
Fax + 352 - 26 29 72 - 39
Email: info.europe@tsrc-global.com
www.tsrcdexco.com

Member of IVK

Company

Year of formation
2011
(for the Europe branch in Luxembourg)

Size of workforce
14

Contact partners
Management:
Christian Kafka

Application technology and sales:
Christine Richter

Range of Products

Types of adhesives
Hot melt adhesives
Solvent-based adhesives
Pressure-sensitive adhesives

Types of sealants
Other (Styrenic Block Copolymers)

Raw materials
Polymers: Styrenic Block Copolymers
(for applications under 1. & 2.)

For applications in the field of
Paper/packaging
Bookbinding/graphic design
Wood/furniture industry
Construction industry, including floors,
walls and ceilings
Automotive industry, aviation industry
Adhesive tapes, labels
Hygiene
Household, recreation and office

Türmerleim AG

Hauptstrasse 15
CH-4102 Binningen
Phone +41 (0) 61 271 21 66
Fax +41 (0) 61 271 21 74
Email: info@tuermerleim.ch
www.tuermerleim.ch

Member of FKS

Company

Year of formation
1992

Size of workforce
8

Managing partners
Marcel Leder-Maeder

Range of Products

Types of adhesives
Hot melt adhesives
Emulsions
Starch, dextrin and casein adhesives
UF-/MUF-resins

For applications in the field of
Paper/packaging
Labelling
Wood/furniture industry
Tissues

Türmerleim GmbH

Arnulfstraße 43
D-67061 Ludwigshafen
Phone +49 (0) 6 21-5 61 07-0
Fax +49 (0) 6 21-5 61 07-12
Email: info@tuermerleim.de
www.tuermerleim.de

Member of IVK, FKS

Company

Year of formation
1889

Size of workforce
130

Managing directors
Matthias Pfeiffer
Dr. Thomas Pfeiffer
Martin Weiland

Subsidiaries
Türmerleim AG, Basel

Contact partners
Management:
Dr. Jörg Liebe
Josef Karl
Tanguy Trippner
Harald Staub

Application technology and sales:
see Management

Range of Products

Types of adhesives
Hot melt adhesives
Emulsions
Starch, dextrin and casein adhesives
UF-/MUF-resins

For applications in the field of
Paper/packaging
Labelling
Wood/furniture industry
Tissues

UHU
GmbH & Co. KG

Herrmannstraße 7
D-77815 Bühl
Phone +49 (0) 72 23-284-0
Fax +49 (0) 72 23-284-288
Email: info@uhu.de
www.UHU.de
www.UHU-profi.de

Member of IVK

Company

Year of formation
1905

Size of workforce
ca. 450 employee

Managing partners
Bolton Group B. V., Amsterdam

Ownership structure
UHU is member of the Bolton Group

Subsidiaries
UHU Austria Ges.m.b.H., Wien (A)
UHU France S.A.R.L., Courbevoie (F)
UHU-BISON Hellas LTD, Pireus (GR)
UHU Ibérica Adesivos, Lda., Lisboa (P)

Sales channels
professional trade, hardware stores,
do-it-yourself hypermarkets, modelbuilding
stores, food trade, stationery, department
stores

Contact partners
Managing Director:
Robert Uytdewillegen, Danny Witjes,
Ralf Schniedenharn

Application technology:
Domenico Verrina

Sales:
Stefan Hilbrath

Range of Products

Types of adhesives
Hot melt adhesives
Reactive adhesives
Solvent-based adhesives
Dispersion adhesives
Pressure-sensitive adhesives

Types of sealants
Acrylic sealants
MS/SMP sealants

For applications in the field of
Paper/packaging
Wood/furniture industry
Construction industry, including floors,
walls and ceilings
Electronics
Mechanical engineering and equipment
construction
Automotive industry, aviation industry
Textile industry
Adhesive tapes, labels
Household, recreation and office

UNITECH Deutschland GmbH

Kaiserstraße 100
D-52134 Herzogenrath
Phone +49 (0) 2407 5 56 90 88
Fax +49 (0) 2407 5 56 90 90
Email: cheolkim@unitech99.co.kr
www.unitech99.co.kr/eng

Member of IVK

Company

Year of formation
1999

Size of workforce
250

Subsidiaries
South Korea (HQ), Slovakia, Germany,
Turkey, China

Sales channels
Direct/Indirect

Contact partners
Management
Dr.rer.nat. Cheol Kim
cheolkim@unitech99.co.kr

Range of Products

Types of adhesives
Reactive adhesives
Epoxy adhesives (1C and 2C solutions)

Types of sealants
PVC-based sealants
Other

For applications in the field of
Automotive industry
(Bodyshop, Paintshop and Interior Shop
materials)
Ship building
Electronics

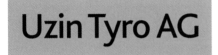

Uzin Tyro AG
Ennetbürgerstrasse 47
CH-6374 Buochs
Phone + 41 41 624 48 88
Fax + 41 41 624 48 89
Email: info@uzin-tyro.ch
www.uzin-tyro.ch

Member of FKS

Company

Year of formation
1933

Size of workforce
38 (55 with subsidiary)

Managing partners
Vitus Meier

Ownership structure
Public company
A company of the Uzin Utz Group
since 1998

Subsidiaries
DS Derendinger AG, Thörishaus, Switzerland

Sales channels
Direct sales, wholesale distribution, trade

Contact partners
Management:
Vitus Meier, Managing Director

Application technology and sales:
Hans Gallati, Head of Sales and Marketing
Switzerland

Further information
Uzin Tyro AG stands for concentrated floor
competence. Since its foundation in the
year 1933, the company has developed
itself to a leading full-range system supplier
for flooring systems in Switzerland. With
the brands UZIN, WOLFF, Pallmann, codex,
RZ, Derendinger and collfox, Uzin Tyro AG
provides a full comprehensive range.

Range of Products

Types of adhesives
Reactive adhesives
Solvent-based adhesives
Dispersion adhesives
Adhesives on special film carrier

Equiment, plant and components
for conveying, mixing, metering and for
adhesive application
for surface pretreatment
measuring and testing

For applications in the field of
Construction industry, including floors,
walls and ceilings
Transport (railway, ships)

Uzin Utz AG

Uzin Utz AG
Dieselstraße 3
D-89079 Ulm
Phone +49 (0) 7 31-40 97-0
Fax +49 (0) 7 31-40 97-1 10
Email: info@uzin-utz.com
www.uzin-utz.com

Member of IVK, FCIO, FKS

Company

Year of formation
1911

Size of workforce
1,060 (December 2016)

Managing partners
Thomas Müllerschön
Beat Ludin
Heinz Leibundgut

Ownership structure
Public company

Subsidiaries
Switzerland, France, Netherland, Belgium,
Great Britain, Poland, Czechia, Austria, USA,
China, Indonesia, New Zealand, Slovenia,
Croatia, Hungary, Serbia, Norway

Sales channels
Trade, Direct marketing

Contact partners
Head of Research and Development
Johanis Tsalos

Head of Sales
Thomas Müllerschön

Further information
Since its foundation in the year 1911, Uzin
Utz AG has developed from a regional
adhesives manufacture to a globally active
full-range system supplier for flooring
systems.

Range of Products

Types of adhesives
Reactive adhesives
Dispersion adhesives
Pressure-sensitive adhesives
Adhesives on PE film carrier

Types of sealants
Acrylic sealants
PUR sealants
Silicone sealants

Equipment, plant and components
for conveying, mixing, metering and
for adhesive application
for surface pretreatment
measuring and testing

For applications in the field of
Construction industry, including floors,
walls and ceilings

Versalis S.p.A.

Piazza Boldrini, 1
20097 San Donato Milanese, Italy
Email: info.elastomers.versalis.eni.com
www.versalis.eni.com

Member of IVK through
Versalis International SA -
Zweigniederlassung Deutschland
Düsseldorfer Straße 13
65760 Eschborn, Germany

Company

Year of formation
1957

Size of workforce
5,200

Managing partners
Ing. Marco Chiappani,
Vice President BU Elastomers

Nominal capital
1.364.790.000 Euro

Ownership structure
Eni S.p.A.

Subsidiaries
see website

Sales channels
Versalis sales network,
see website

Contact partners
Management:
Ing. Franco Ossola,
Sales Manager Elastomers + local sales
network in single countries

Further information
See official website:
www.versalis.eni.com

Range of Products

Types of adhesives
Hot melt adhesives
Solvent-based adhesives
Pressure-sensitive adhesives

Raw materials
Polymers

For applications in the field of
Paper/packaging
Bookbinding/graphic design
Wood/furniture industry
Automotive industry, aviation industry
Adhesive tapes, labels
Hygiene
Household, recreation and office

Vinavil S. p. A.

Via Valtellina, 63
I-20159 Milano
Phone +39-02-69 55 41
Fax +39-02-69 55 48 90
Email: vinavil@vinavil.it
www.vinavil.it

Member of IVK

Company

Year of formation
1994

Size of workforce
> 300

Ownership structure
Mapei S. p. A.

Subsidiaries
Vinavil Americas. Corp.
Vinavil Egypt

Sales channels
> 40 representative commercial offices

Contact partners
Management:
Taako Brouwer
Ing. Silvio Pellerani

Application technology and sales:
Ing. Silvio Pellerani
Dr. Marco Cerra
Dr. Fabio Chiozza

Further information
Certified acc. to ISO EN 9001, 14001 and
OHSAS 18001

Range of Products

Types of adhesives
Dispersion adhesives
Pressure-sensitive adhesives

Types of sealants
Acrylic sealants

Raw materials
Polymers:
aqueous polymer dispersions,
solid resins and redispersible powders
RAVEMUL®, VINAVIL®, CRILAT®, RAVIFLEX®
and VINAFLEX® based on vinylacetate, vinylacetate copolymers, vinylacetate ethylene
copolymers, acrylic and styrene acrylic

For applications in the field of
Paper/packaging
Bookbinding/graphic design
Wood/furniture industry
Construction industry, including floors,
walls and ceilings
Automotive industry, aviation industry
Textile industry
Adhesive tapes, labels
Household, recreation and office

VITO IRMEN GmbH & Co. KG
Mittelstraße 74 – 80
D-53424 Remagen
Phone +49 (0) 26 42-4 00 70
Fax +49 (0) 26 42-4 29 13
Email: info@vito-irmen.de
www.vito-irmen.de

Member of IVK

Company

Year of formation
1907

Size of workforce
85

Managing director
Ralf Heiligtag

Nominal capital
4,000,000 €

Ownership structure
Limited commercial partnership

Subsidiaries
Representatives in Poland, Austria, Spain, Netherlands, Czech Republic, Hungary, Russia

Sales channels
direct and via dealers & distributors

Contact partners
Management:
Ralf Heiligtag

Application technology and sales:
Erich Dochow

Further information
www.vito-irmen.de

Range of Products

Self-adhesives tapes coated with
Hot melt adhesives
Solvent-based adhesives
Dispersion adhesives
Pressure-sensitive adhesives

For applications in the field of
- Adhesive tapes, labels
- Automotive industry/aviation industry
- Construction industry, including floors, walls and ceilings
- Device for production, storage and transport of glass
- Electronic Assembly
- Hygiene
- Mechanical engineering and equipment construction
- Solar Industry
- Structural Glazing
- Wood/furniture industry

Wacker Chemie AG
Hanns-Seidel-Platz 4
D-81737 Munich
Phone +49 (0) 89-62 79-17 41
Fax +49 (0) 89-62 79-17 70
Email: info@wacker.com
www.wacker.com

Member of IVK

Company

Year of foundation
1914

Size of workforce
ca. 17,000

Ownership structure
Stock corporation ("Aktiengesellschaft")

Subsidiaries
25 production sites, 21 technical competence centers and 52 sales offices globally.

Range of Products

Raw Materials
Vinyl acetate polymers:
dispersions, dispersible polymer powders
and solid resins (VINNAPAS® and VINNEX®)
Vinyl acetate/ethylene copolymers:
dispersions and dispersible polymer
powders (VINNAPAS® and VINNEX®)
VC copolymers (VINNOL®)
Silicones

Additives:
Pyrogenic silica (HDK®)
Silanes, adhesion promoters and cross-
linkers (GENIOSIL®)
Foam-control agents and silicone
surfactants
Nanoscale silicone particles for modifying
adhesives (GENIOPERL®)

Sealants and Adhesive Grades
RTV-1 Silicones (ELASTOSIL®)
RTV-2 Silicones
LSR Silicones
Silicone gels and silicone foams
UV-curing systems
Hybrid adhesives (GENIOSIL®)

Wakol GmbH

Bottenbacher Straße 30
D-66954 Pirmasens
Phone +49 (0) 63 31-80 01-0
Email: info@wakol.com
www.wakol.com

Member of IVK, FCIO, FKS

Company

Year of formation
1934

Size of workforce
210

Management
Steffen Acker
Christian Groß (CEO)
Dr. Frederic Holzbaur
Dr. Martin Schäfer

Subsidiaries
Austria, Switzerland, Poland, Italy, USA

Sales channels
direct distribution, specialised trade

Range of Products

Types of adhesives
Dispersion-based adhesives &
Sealing Compounds
Reactive adhesives
Solvent-based adhesives
Hot melt adhesives
PVC Plastisols
PU Foam

For applications in the field of
Construction industry, including floors,
walls and ceilings
Automotive industry, aviation industry,
including passenger seats,
Wood/furniture industry, Metal packaging
industry, Footwear & Leather industry

WEICON GmbH & Co. KG
Koenigsberger Straße 255
D-48157 Muenster
Phone +49 (0) 2 51-93 22-0
Fax +49 (0) 2 51-93 22-2 44
E-mail: info@weicon.de
www.weicon.de

Member of IVK

Company

Year of formation
1947

Size of workforce
210

Subsidiaries
WEICON Middle East L.L.C
WEICON Kimya Sanayi Tic. Ltd. Sti.
WEICON Inc.
WEICON Romania SRL
WEICON SA (Pty) Ltd
WEICON South East Asia Pte Ltd
WEICON Czech Republic s.r.o.
WEICON Iberica S.L.

Sales channels
Technical Distributors, Industry directly

Contact partners
Management:
Ralph Weidling
Timo Gratilow

Application technology and sales:
Holger Lütfring
Product Manager

Vitali Walter
Export Manager

Range of Products

Types of adhesives
2-comp. adhesives
Basis: Epoxy resins, PUR, MMA
1-comp. adhesives
Basis: Cyanoacrylate, PUR, MMA, POP
Reactive adhesives
Solvent-based adhesives

Types of sealants
PUR-Sealants
Silicone sealants
MS/SMP-Sealants

For applications in the field of
Paper/packaging
Wood/furniture industry
Construction industry, including floors, walls
and ceilings
Electronics
Mechanical engineering and equipment
construction
Automotive industry, aviation industry
Household, recreation and office

**Weiss Chemie + Technik
GmbH & Co. KG**
Hansastraße 2
D-35708 Haiger
Phone +49 (0) 27 73-8 15-0
Fax +49 (0) 27 73-8 15-2 00
Email: ks@weiss-chemie.de
www.weiss-chemie.de

Member of IVK, GEV

Company

Year of formation
1815

Size of workforce
300 members of staff within the group

Managing partners
WBV – Weiss Beteiligungs- und Verwaltungs-
gesellschaft mbH

Subsidiaries
Haiger, Herzebrock, Niederdreisbach,
Monroe NC (USA)

Nominal capital
2 Mio. €

Ownership structure
Family share holders

Management
Jürgen Grimm

Contact partners
Reception: Phone +49 (0) 27 73-8 15-0
Sales: Phone +49 (0) 27 73-8 15-2 02
Technology: Phone +49 (0) 27 73-8 15-2 55
Purchase: Phone +49 (0) 27 73-8 15-2 41

Sales channels worldwide
Own sales force, specialist distributors
industry and trade, customers with private
label

Range of Products

Business division adhesives

Types of adhesives
Reactive adhesives (PUR, Epoxy)
Cyanoacrylate instant glues
Hybrid adhesives (STP/MS)
Solvent-based adhesives
Dispersion adhesives
Pressure sensitive adhesives
Cleaners, solvent- and tenside based

For applications in the field of
Window- and door industry
(plastic materials, metal, wood)
Airtight building envelope according to EnEV
Dry construction
Transportation/commercial vehicles,
Shipbuilding, railway vehicles,
Caravan industry, container construction
Fire protection
Air-conditioning technology and ventilation
engineering
Sandwich- and composite panels
Building trade
Wood-/furniture industry

Business division sandwich elements
Light sandwich constructions as heat- and
sound insulating sandwich elements applied
in fields like doors, windows, gates, booth
constructions, automotive industry etc.

Willers, Engel & Co. GmbH

Lippeltstraße 1
D-20097 Hamburg
Phone +49 (0) 40 33 7967
Fax +49 (0) 40 33 1980
Email: office@willersengel.de
www.drt-france.com

Member of IVK

Company

Year of formation
1910

Size of workforce
8

Nominal capital
400,000 €

Ownership structure
100 % subsidiary of D.R.T.

Contact partners
Management:
Jens Döhle

Application technology and sales:
Lars-Olaf Jessen
Rüdiger Heidorn

Further information
Sales of:
Rosin, Rosin Derivatives, Resin Dispersions
Terpene Phenolic Resins, Polyterpene Resins
Terpenes

Range of Products

Raw materials
Resins
Solvents

For applications in the field of
Paper/packaging
Bookbinding/graphic design
Wood/furniture industry
Construction industry, including floors, walls and ceilings
Electronics
Mechanical enineering and equipment construction
Automotive industry, aviation industry
Textile industry
Adhesive tapes, labels
Hygiene
Household, recreation and office

Worlée-Chemie GmbH
Grusonstraße 22
D–22113 Hamburg
Phone +49 (0) 40-7 33 33-0
Fax +49 (0) 40-7 33 33-11 70
Email: service@worlee.de
www.worlee.com

Member of IVK

Company

Year established
1962
(founding as subsidary of E.H.Worlée & Co.
established 1851 as trading company)

Employees
about 250
(productions sites in Lauenburg and Lubeck,
regional sales offices in Germany and affiliates
abroad)

Share holder
Dr. Albrecht von Eben-Worlée
Reinhold von Eben-Worlée

Properties/Estate
own by the von Eben-Worlée family
Affiliates:
E.H. Worlée & Co. b.V. Kortenhoef (NL)
E. H. Worlée & Co. (UK) Ltd.
Newcastel-under-Lyme (GB)
Worlée Chemie India Private Limited,
Mumbai(India)
Worlée Italia S.R.L. Mailand (I)
Varistor AG, Neuenhof (CH)
Worlée (Shanghai) Trading Co. Ltd.
Shanghai (CN)

Sales Organisation
regional sales organisations in Germany,
affiliates, distributors and agents abroad

Management
Dr. Albrecht von Eben-Worlée
Reinhold von Eben-Worlée
Joachim Freude

Range of Products

Additives
Acrylic resins
Acrylic dispersions
Alkyd resins
Alkyd emulsions
Polyester
Polyester polyols
Maleic resins
Rosin based hard resins, phenol modified
Adhesion promoter
Special Primer
Pigments
XSBR- water based dispersions of carboxylated
styrene-butadiene latex
HS-SBR water based dispersion of
styrene-butadiene copolymer (high solid)
VA – Vinyl acetate dispersions
VAA – Vinyl acetate copolymer dispersions
PVAC – Polyvinyl acetate resins
Styrene acrylic resins
Butadien acrylic resins
Thickeners
Aliphatic isocyanates
Polyisocyanates
Polythiols
Hydro carbon binder resins

WULFF GmbH
u. Co. KG

Wersener Straße 3
D-49504 Lotte
Phone +49 (0) 5404-881-0
Fax +49 (0) 5404-881-849
Email: industrie@wulff-gmbh.de
www.wulff-gmbh.de

Member of IVK

Company

Year of formation
1890

Size of workforce
180

Managing partners
Familiy Israel, Mr. Ernst Dieckmann

Sales channels
directly and handling

Contact partners
Management:
Mr. Alexander Israel, Mr. Jan-Steffen Entrup

Application technology and sales:
Mr. Ralf Hummelt, Dr. Michael Erberich,
Mr. Jörg Dronia

Range of Products

Types of adhesives
Dispersion adhesives

Types of sealants
MS/SMP sealants

For applications in the field of
Construction industry, including floors,
walls and ceilings
Textile industry

Zelu Chemie GmbH
Robert-Bosch-Straße 8
D-71711 Murr
Phone +49 (0) 71 44-82 57-0
Fax +49 (0) 71 44-82 57-30
Email: info@zelu.de
www.zelu.de

Member of IVK

Company

Year of formation
1889

Size of workforce
~ 50 employees

Ownership structure
GmbH

Sales channels
Technical Sales
International Sales representatives

R&D and Application Development:
Dr. Stefan Kissling
Team Leader Adhesives
Phone: +49 (0) 71 44-82 57-23
Email: s.kissling@zelu.de

Range of Products

Types of adhesives
Hotmelts
Water-based dispersion adhesives
Solvent-based adhesives
Reactive two-component PUR

For applications in the field of
Foam converting industry
Furniture-/Upholstery industry
Mattress production
Automotive interior laminations
Filter production

Equipment and Plant Manufacturing

)(BÜHNEN

Bühnen GmbH & Co. KG
Hinterm Sielhof 25
D-28277 Bremen
Phone +49 (0) 4 21-51 20-0
Fax +49 (0) 4 21-51 20-2 60
Email: info@buehnen.de
www.buehnen.de

Member of IVK

Company

Year of formation
1922

Size of workforce
89

Ownership structure
Private ownership

Subsidiaries
BÜHNEN Polska Sp. z o.o.
BÜHNEN B.V., NL
BÜHNEN, AT

Contact person
Managing Director:
Bert Gausepohl

International Sales/Marketing
Valentino di Candido

Sales GER, AT, CH
Hans-Gerhard Hartje

Distribution channels
Direct Sales, Distributors

Range of Products

Hot Melt Adhesives
The product range includes a
variety of different hot melt adhesives
for almost every application.
Available bases:
EVA, PO, POR, PA, PSA, PUR, Acrylate.
Available shapes:
slugs, sticks, granules, pillows, blocks,
cartridges, barrels, drums, bags.

Application Technology
Hot melt tank applicator systems with
piston pump or gear pump, PUR- and POR-
hot melt tank systems, PUR- and POR-bulk
unloader, hand guns for spray and line
application, application heads for line, slot,
spray, dot, spiral application and special
application heads for individual customer
requirements, hand-operated glue applicators, PUR- and POR glue applicators, wide
range of application accessories, customer-
oriented application, solutions.

Applications
Automotive, Packaging, Display Manufac-
turing, Electronic Industry, Filter Industry,
Shoe Industry, Foamplastic and Textile
Industry, Case Industry, Construction
Industry, Florists, Wood-Processing and
Furniture Industry, labelling Industry.

Drei Bond GmbH
Carl-Zeiss-Ring 17
85737 Ismaning, Germany
Phone +49 (0) 89-962427 0
Fax +49 (0) 89-962427 19
Email: info@dreibond.de
www.dreibond.de

Member of IVK

Company

Year of formation
1979

Number of employees
48

Partners
Drei Bond Holding GmbH

Share capital
€ 50,618

Subsidiaries
Drei Bond Polska sp. z o.o. in Kraków

Distribution channels
Directly to the automotive industry
(OEM + tier 1/tier 2); indirectly
via trading partners as well as select private
label business

Contacts
Management:
Mr. Thomas Brandl

Application engineering, adhesive and
sealants:
Christian Eicke

Application engineering, metering technology:
Sebastian Schmid, Norbert Frank,
Marco Hein

Adhesive and sealant sales:
Christian Eicke

Metering technology sales:
Norbert Frank, Marco Hein

Additional information
Drei Bond is certified according to ISO 9001-
2015 and ISO 14001-2015

Range of Products

Types of adhesives/sealants
• Cyanoacrylate adhesives
• Anaerobic adhesives and sealants
• UV-light curing adhesives
• 1C/2C epoxy adhesives
• 2C MMA adhesives
• 1C/2C PUR adhesives
• 1C MS hybrid adhesives and sealants
• 1C synthetic adhesives and sealants
• 1C silicone sealants

Complementary products:
• Activators, primers, cleaners

Equipment, systems and components
• Drei Bond Compact metering systems →
  semi-automatic application of adhesives and
  sealants, greases and oils
  Metering technology: pressure/time and
  volumetric
• Drei Bond Inline metering systems →fully
  automated application of adhesives and
  sealants, greases and oils
  Metering technology: pressure/time and
  volumetric
• Drei Bond metering components:
  Container systems: tanks, cartridges, drum
  pumps
  Metering valves: progressive cavity pumps,
  diaphragm valves, pinch valves, spray
  valves, rotor spray

For applications in the following fields
• Automotive industry/automotive suppliers
• Electronics industry
• Elastomer/plastics/metal processing
• Mechanical and apparatus engineering
• Engine and gear manufacturing
• Enclosure manufacturing (metal and plastic)

Gößl + Pfaff GmbH

Münchener Straße 13
D-85123 Karlskron
Phone +49 8450-9320
Fax +49 8450-932 13
Email: info@goessel-pfaff.de
www.goessl-pfaff.de

Member of IVK

Company

Year of formation
1984

Size of workforce
23

Managing partners
Roland Gößl
Josef Pfaff

Sales channels
Technischer Vertrieb/Web-Shop

Contact partners
Management:
Roland Gößl
Josef Pfaff

Application technology and sales:
Kathrin Pfaff
Martina Reithmeier

Range of Products

Types of adhesives
Reactive adhesives

Equipment, plant and components
for conveying, mixing, metering and for
adhesive application

For applications in the field of
Wood/furniture industry
Construction industry, including floors, walls
and ceilings
Electronics
Mechanical engineering and equipment
construction
Automotive industry, aviation industry

GRACO Distribution BVBA
European Headquarters
Slakweidestraat 31
B-3630 Maasmechelen, Belgium
Phone +32 89 770 700
Fax +32 89 770 700
Email: communications@graco.be
www.graco.com

Company

Year of formation
1926

Size of workforce worldwide
2,600

Sales channels
Specialized distributors partners worldwide

Contact partners
Public relations:
Miranda Houbrechts

Trade Marketing Specialist:
Chris Li Citra

Further information
Every day around the world, you'll find Graco. It's an airplane in the U.S., a windmill in Brussels, a railroad in Shanghai, a sports arena in South America. We have painted, finished, filled, glued and sealed components in all of these things and more.
We pump the peanut butter and mayonnaise into your jar, the filling into your cookie, the oil into your car. We glue the sole of your shoe and pump the ink onto your currency. We spray varnish on your furniture, coating on your pills, the paint on your house and texture on your ceiling. We even sprayed the paint on the United States White House and stripes on the field at Wembley Stadium in London, England.
We're leading fluid handling company, marketing products in more than 100 countries on six continents.

Range of Products

Most common materials for the sealant and adhesive market
All sealants & adhesives: also includes acrylates, hot melts, silicones, epoxies, polyester paints and gel coats.

Types of machines
Single component and multi component continuous output

For applications in the field of
Transport, Packaging, Glass, Electronics, Alternative Energy and more ...

Infinite possibilities for endless applications

Bonding and Sealing Equipment

Discover our great range of equipment for all your sealing and bonding applications. Whether you're working with 1k, 2k or hot melt material, we can provide you the equipment that will fulfill your needs.

Discover your Solution

www.graco-sae.com

HARDO-Maschinenbau GmbH
Grüner Sand 78
32107 Bad Salzuflen
Germany
Phone +49 (0) 5222-9301-5
Fax +49 (0) 5222-93016
Email: thermo@hardo.eu
www.hardo.eu

Company

Year of formation
1935

Size of workforce
70 employees

Ownership structure
Family owned

Nominal capital
1.022.500 €

Sales channels
Direct sales and distribution partners
worldwide

Contact partners
Managing director:
Ind. & Econ. Engineer Ingo Hausdorf

Sales management:
Hauke Michael Immig

Sales and application technology:
Carsten Schoeler
Ralf Drexhage
Thomas Sonnenberg
Reinhard Kölling

Further information
More than 50 years experience in developing
and manufacturing of adhesive application
and premelting solutions.

Range of Products

Application systems for
Hotmelts
Pressure sensitive hotmelts
Reactive hotmelts
Dispersion adhesives
Reactive adhesives
Water based primers
Butyl
Bituminous masses
Waxes
Diverse substances

Premelting systems for
Hotmelts
Reactive hotmelts

Roller laminating systems
Tape presses
Plate presses
Winding systems

Customer-specific solutions
HARDO is your ideal partner when it comes
to customer-specific solutions. Our team
of application technicians and engineers
will work together with you to create the
optimum machine or system for your work
process. Our laboratory will identify the
best possible application system for your
particular case.

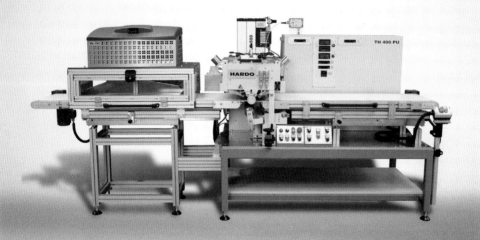

Dr. Hönle AG – UV-Technology
Head of Hönle Group
Lochhamer Schlag 1
D-82166 Gräfelfing/München
Phone +49 (0) 89-85 60 80
Fax +49 (0) 89-85 60 81 48
Email: uv@hoenle.de
www.hoenle.de

Company

Year of formation
1976

Annual turnover
93,4 Mio. Euro

Managing Board
Norbert Haimerl
Heiko Runge

Tochterfirmen im Klebstoffbereich
D: Panacol-Elosol GmbH Deutschland
     (Adhesives) – Steinbach/Taunus
     Aladin GmbH (UV lamp manufacturing) –
     Rott am Inn
F: Eleco Produits S.A.S. – F-Gennevillers
     Cedex
I: Hönle Italy, Sales Office
US: Tangent Industries Inc. – US-Torrington
KOR: SKC-Panacol Co., Ltd. – KOR-Suwon-si

Contact Partners
Sales Director: Dieter Stirner
Team Leader/Sales: Florian Diermeier

Sales channels
Sales team for Germany
International distribution network with world-
wide sales partners

Range of Products

UV/UV-LED technology
for curing UV reactive adhesives,
sealing & potting materials
and plastics
for drying and curing UV reactive inks
and coatings
for fluorescence testing
for sun simulation

For applications in the field of
Electronics and microelectronics
manufacturing
Conformal coating/chip encapsulation
Precision mechanics
Plants and machinery manufacturing
Automobile and aerospace industry
Glass industry
Optics
Medical engineering
Quality control

Adhesives
See Panacol, page 105

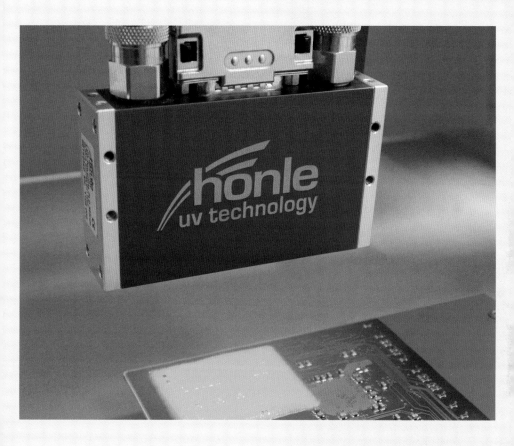

**Innotech Marketing und
Konfektion Rot GmbH**
Schönbornstraße 8
D-69242 Rettigheim
Phone +49 (0) 7253-98 88 55 50
Fax +49 (0) 7253-932 40 77
Email: verkauf@innotech-rot.de
www.innotech-rot.de

Member of IVK

Company

Year of formation
1995

Size of workforce
22

Managing partners
Joachim Rapp, Anja Gaber

Nominal capital
100.000 €

Ownership structure
GmbH

Subsidiaries
Adhetek GmbH

Sales channels
technical trade, direct sales, independent sales
representatives, trade shows

Contact partners
Management: Joachim Rapp, Anja Gaber

Technology/Application equipment: Martin Deutsch

Application Solutions/Adhesive Accessories:
Nadine Knörr

Further information
Innotech offers a wide range and expertise in the
field of bonding and sealing. The company is
specialized on applicators and adhesive accessories from all leading manufacturers. Competent
support & advice, trade, training courses, marketing samples, consulting and repair service
completes Innotech's wide range and expertise in
bonding and sealing. New at Innotech: Bonding
trainings as an official cooperation partner of
Fraunhofer IFAM for European Adhesive Bonder
(EAB) and European Adhesive Specialist (EAS).

Range of Products

Applicators
As distributor and service partner of international leading sealant and adhesive manufacturers, Innotech offers more than 600 models
of applicators and dispensing tools – quality
made in Europe and USA- with more than
200.000 spare parts. This involves the repair
and warranty service as well as the distribution of specialized tools, just-in-time delivery,
consumer parts and accessories.

Adhesive accessories
Wide assortment and worldwide dispatch of
mixers and nozzles and surface treatment
from all leading manufacturers.

Services
Support and advice for applicators, Repair service
for applicators, Support, solutions and training,
Cooperation partner with IFAM – Training Center
for European Adhesive Specialist (EAS) and
European Adhesive Bonder (EAB), Conduction of
training sessions of European Adhesion Specialist
(EAS) & European Adhesive Bonder (EAB),
Logistic and transportation, Sample logistics,
Development of special / tailormade nozzles,
Product innovations

Trade
Adhesives, Special applicators, Cartridges
Mixers and nozzles, Adhesive supplies, Cleaners,
Primer equipment, Surface treatment Pyrosil

Production
Marketing bonding, Refilling service
(in different quantities), Test pieces, Packaging/
Assembling

IST METZ GmbH
Lauterstraße 14 – 18
D-72622 Nürtingen
Phone +49 (0) 0 70 22 - 6 00 20
Fax +49 (0) 0 70 22 - 6 00 276
Email: info@ist-uv.com
www.ist-uv.com

Member of IVK

Company

Year of formation
1977

Size of workforce
500 employees worldwide

Ownership structure
GmbH

Subsidiaries
eta plus electronic GmbH
Integration Technology Ltd.
IST France sarl
IST Italia S.r.l.
IST (UK) Limited
IST Nordic AB
IST Benelux B.V.
UV-IST Ibérica SL
IST America Corp.
IST METZ SEA Co., Ltd.
IST METZ UV Equipment China Ltd. Co.
IST East Asia K.K.

Contact partners
Management:
Christian-Marius Metz

Application technology and sales:
Arnd Riekenbrauck

Range of Products

Equipment, plant and components
for adhesive curing
adhesive curing and drying
measuring and testing

For applications in the field of
Paper / packaging
Building industry
Construction industry, including floors,
walls and ceilings
Electronics
Automotive industry, aviation industry
Adhesive tapes, labels

Nordson Deutschland GmbH
Heinrich-Hertz-Straße 42
D-40699 Erkrath
Phone +49 (0) 211 92 05-0
Fax +49 (0) 211 25 46 58
Email: info@de.nordson.com
www.nordson.de

Company

Year of formation
1967

Size of the workforce
450 employees with Nordson Engineering
(Lüneburg)

Shareholders
Nordson Corporation, USA

Contact partners
Board of management:
Axel Wenz, Ulrich Bender und
Srinivas Subramanian

Sales
General sales manager:
Georg Gillessen
OEM support:
Christian Schwär
Packaging/assembly applications:
Georg Gillessen

Industrial applications: Jörg Klein
Nonwoven: Kai Kröger
Industrial Coating Systems: Michael Lazin
Container: Ralf Scheuffgen
Automotive: Volker Jagielki

Sales routes
Through field service staff of Nordson
Deutschland GmbH

Further information
Development centres and production facilities
(ISO-certified) in the USA and Europe, over 7,200
employees, subsidiaries on every continent. In
cooperation with the customer, Nordson develops
complete solutions with integrated systems
and matching components which grow with the
customers' demands.

Range of Products

Equipment and systems for the application of
adhesives and sealants and for surface finishing
with paints, lacquers, other liquid materials or
powders. Nordson systems can be integrated into
existing production lines.

Packaging and assembly applications
Complete adhesive application systems (hot
melt/cold adhesive) for integration into packaging lines. In the field of assembly applications,
Nordson optimises production processes in many
different branches of industry.

Industrial applications
Adhesive and sealant applications for a wide
variety of branches of industry, e.g. filter bonding,
sealing of vehicle rear lights and mattress bonding. Wood processing, e.g. profile wrapping, edge
banding and post-forming.

Nonwoven
Nonwoven (tailored systems for the application
of adhesive and super-absorbent powder for the
production of babies' nappies, slip inlays, sanitary
towels and incontinence articles).

Powder & Liquid Coating
Applications and systems for coating with paints,
lacquers, other liquid materials and powders.

Containers
Applications and systems for coating and labelling
cans, containers and other vessels.

Electronics
Automatic coating and metering plants for the
electronics industry for precise application of
adhesives, grouting compounds, solder pastes,
fluxes, protective lacquers, etc.

Automotive
Engineered systems for the application of structural adhesives and sealants in harsh automotive
manufacturing environments.

Reinhardt-Technik GmbH
a Member of WAGNER GROUP
Waldheimstraße 3
D-58566 Kierspe
Phone +49 (0) 2359 666-0
Fax +49 (0) 2359 666-129
Email: info-rt@wagner-group.com
www.reinhardt-technik.com

Company

Year of formation
1962

Size of workforce
About 100 employees

CEO
Christian Glaser

Sales channels
Sales representatives, agents and distributors

Company Profile
Reinhardt-Technik GmbH specializes in the areas of bonding, sealing and casting including injection molding. The company offers a comprehensive range of machines for metering and mixing technology, which process cold or heated 1K materials and multi-component liquid plastics such as polyurethanes, polysulphides, epoxies, silicones and LSR (Liquid Silicone Rubber). All common process technologies are offered - from pneumatically or hydraulically driven piston pumps, gear metering systems to electrically controlled shot meter systems as well as progressive cavity metering units. In addition, Reinhardt-Technik is a total solution provider and supplies complete custom-engineered manufacturing cells with robots, coupled with handling and inspection systems as stand-alone solutions or integrated into a production line.

Range of Products

Metering and Potting Systems
Highly precise and reliable dosing systems for a variety of applications:
• Shot meter systems
• Gear metering systems
• Progressive cavity metering systems

Automated Production Cells
Convenient, simple as well as safe operation of the dosing and mixing systems along with integration with upstream and downstream processes:
• Standardized robot cells
• Individual and application-specific manufacturing cells

Material Preparation and Feeding
Modular as well as configurable processing and conveying systems:
• Agitators for fast-sedimenting materials
• 1K- or 2K-compents
• Optional booster addition
• 20 and 200 litre feeding units with robust working cylinder in combination with powerful chop check pumps for low viscosity materials
• Optional pressure containers
• Reliable material discharge with low residual material

Mixing Systems
Low-maintenance and powerful mixing systems for individual processing:
• Static mixing systems
• Dynamic mixing systems
• Incl. snuff back valve

Customer Service
• Technical application centre
• Training
• Hotline
• Remote maintenance
• Online shop (spare and wearing parts)
• High spare part availability
• Service worldwide
• Equipment modernization and modification
• Process optimization

Robatech
Gluing Technology

Robatech AG
Pilatusring 10
CH-5630 Muri AG
Phone (+41) 56 675 77-00
Fax (+41) 56 675 77-01
Email: info@robatech.ch
www.robatech.ch

Member of IVK

Company

Year of formation
1975

Size of workforce
More than 600 employees all over the world

Shareholder
Robatech AG, CH-5630 Muri, Switzerland

Ownership structure
Robatech AG, CH-5630 Muri, Switzerland

Subsidiaries and agencies
Represented in more than 60 countries worldwide
Germany: Robatech GmbH,
Im Gründchen 2, D-65520 Bad Camberg
Phone +49 (0) 64 34-94 11-0
Fax +49 (0) 64 34-94 11-22

Channel of distribuition
Via Head office, subsidiaries and agencies

Contact partners
Management:
Robatech AG, Switzerland:
Marcel Meyer
Robatech GmbH, Germany:
Eberhard Schlicht, Andreas Schmidt

Application technology and sales:
Robatech AG, Switzerland:
Kishor Butani, Sales Director
Kevin Ahlers, Marketing Director
Robatech GmbH, Germany:
Eberhard Schlicht, Managing Director

Further information
Production facility in Germany, Hong Kong and Switzerland

Range of Products

Product and sales program of the company
- Glue application system with piston pumps and gear pumps for hotmelt and dispersions, inclusive necessary equipment.
- Small hotmelt application systems up to 5 liter tank capacity
- Medium hotmelt application systems from 5 to 30 liter tank capacity
- Big hotmelt application systems from 55 to 160 liter tank capacity
- Hotmelt application systems for PUR-hotmelts from 3 to 30 liter tank capacity
- Drumunloaders from 50 to 200 liter tank capacity
- Application technics: Bead application, Surface coating, Spray application, spiral application
- Electrical timing and electrical metering
- Cold glue application systems: Pressure tanks and pump systems

Robatech offers a solution for several industries
Packaging Industry
Converting Industry
Graphic Industry
Hygienic Industry
Textile-converting Industry
Woodworking Industry
Building Supplies Industry
Automotive Industry
Various Industries

GLUING
WITH PERFECTION

STREAM. THE NEW CONCEPT.

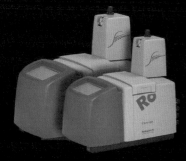

- innovative melting technology (tankless)
- fast heat-up time
- increased operator safety

Bring flow into your gluing process.

More info
robatech.com/
concept-stream

CONCEPT *Stream*

Robatech
Gluing Technology

Part of the Atlas Copco Group

SCA Schucker GmbH & Co. KG
Gewerbestrasse 52
D-75015 Bretten
Phone +49 (0) 7252 5560-0
Fax +49 (0) 7252 5560-51 00
Email: info@sca-solutions.com
www.sca-solutions.com

Company

Year of establishment
1986. Part of the Atlas Copco Group since 2011

Workforce
More than 650 employees

Shareholders
Atlas Copco Holding GmbH, Essen

Subsidiaries
Affiliates and service offices in more than
27 countries
You will find contacts throughout the world at
www.sca-solutions.com

Sales channels
Direct Sales

Contact partners
Board of Management:
Olaf Leonhardt and Dieter Eltschkner,
General Managers of SCA Schucker GmbH & Co. KG

Further information
SCA develops and produces systems and equipment
for the application of adhesives and sealants. From
the adhesive pump through to the application nozzle,
the product range includes all the components required for handling an adhesive or sealant, for controlled adhesive application, for interchange with the
robot, for system diagnosis and for process data
storage. SCA has a reputation for excellent quality,
especially in the automotive industry. As it is part of
the Atlas Copco Group, SCA can solve complex
joining tasks by combining several joining technologies. Atlas Copco is a world leader in the joining
technologies of adhesive bonding (SCA), bolting
(Atlas Copco Tools) and self-pierce riveting (Henrob).
Competence in other joining technologies such as
welding or flow drill screw driving is provided by
partner companies. SCA combines classic joining
technologies with adhesive bonding in such a way
as to exploit the benefits of the individual technologies while avoiding their drawbacks. Customers
benefit from comprehensive process know how
within the Group.

Range of Products

**Processing of the following types of adhesives
and sealants**
- Acrylic
- Butyl
- Polyurethane
- Silicone
- (Modified) polymer
- PVC
- Epoxy resin
- Water-based acrylate

Systems/processes/accessories/services
- Application systems (single-component and
  2-component) for adhesives and sealants
- Pumping, mixing and metering components
- Surface treatment
- Measurement and testing systems
- Robot cell solutions for small-series production
- Hybrid joining technologies
- Simulation of new adhesives and processes
- Comprehensive world-wide service: initial tests,
  project planning, project management, instal-
  lation and commissioning, maintenance, repair,
  preventive maintenance, process optimization,
  training etc.

For applications in the field of
- Automotive industry
- Aerospace industry
- Transport industry
- Mechanical engineering
- Domestic equipment and appliances
- Electric vehicles and electric powertrain

 Scheugenpflug

Scheugenpflug AG
Gewerbepark 23
D-93333 Neustadt
Phone +49 (0) 94 45-95 64-0
Fax +49 (0) 94 45-95 64-40
Email: vertrieb.de@scheugenpflug.de
www.scheugenpflug.de

Company

Year of formation
1990

Size of workforce
More than 450 (worldwide)

Managing Board
Johann Gerneth, Christian Ostermeier

Contact
Email: vertrieb.de@scheugenpflug.de
www.scheugenpflug.de

Subsidiaries
Scheugenpflug Resin Metering Technologies Co.,
Ltd., China
Email: info@scheugenpflug.com.cn

Scheugenpflug Inc., USA
Email: sales.usa@scheugenpflug-usa.com

Scheugenpflug México, S. de R. L. de C. V.
Email: sales.mx@scheugenpflug-usa.com

Sales Partners Worldwide
See Webpage > Company > Locations & Sales
Partners

Company Profile
Technology that sets standards: With over 25 years
of experience, Scheugenpflug is one of the leading
manufacturers of innovative adhesive bonding, dispensing and potting technology. With an additional core
competency in process automation, the product and
technology range extends from cutting-edge material
preparation and feeding units and high performance
dispensing systems to custom-tailored, modular
production lines for a wide range of applications.
Scheugenpflug systems are used in the automotive
and electronics industries as well as the telecommuni-
cations sector, medical technology and the chemical
industry. The company has four additional locations in
the USA, China (2x) and Mexico as well as numerous
service locations and sales partners all over the world.

Range of Products

Dispensers
• Volumetric piston dispensers
• Alternating volumetric piston dispensers
• A volumetric piston dispenser specifically for
  thermally conductive materials
• Gear pump dispensers
• Small quantity and micro dispensers

Material Preparation and Feeding
• Feeding from cartridges and pressure tanks
• Feeding from hobbocks
• Systems for self-leveling potting media

Dispensing Systems
• Manual work stations with stand or for
  integration
• Dispensing cells
• Processing modules for integration

Vacuum Potting Systems
• Entry-level systems
• Large vacuum chambers

Customized Systems and Solutions
Customized automated solutions for various
adhesive bonding, dispensing and potting tasks,
based on the Scheugenpflug modular system
(scalable design)

Services
• Technology and Application Center/
  Dispensing tests
• Technical support/Maintenance/Hotline
• After Sales/Spare parts
• Trainings
• Rental equipment
• Subcontracting
• Interactive training videos for system operators
  (SISS)

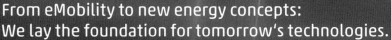

Sonderhoff Engineering GmbH
Dr. Walter Zumtobel Straße 15
A-6850 Dornbirn
Phone: +43 5572 39 88 10
Fax: +43 5572 39 88 10-55
Email: info@sonderhoff.com
www.sonderhoff.com

Company

Year of formation
1988

Size of workforce
94 employees

More than 260 employees within Sonderhoff Group of Companies worldwide

Ownership structure
Sonderhoff Holding GmbH

Sister companies
Sonderhoff Chemicals GmbH, Cologne (Germany)
Sonderhoff Services GmbH, Cologne (Germany)
Sonderhoff Polymer-Services Austria GmbH, Hörbranz (Austria)
Sonderhoff Italia SRL, Oggiono (Italy)
Sonderhoff USA Corporation, Elgin (USA)
Sonderhoff (Suzhou) Sealing Systems Co.Ltd, Suzhou (China)

Sales channels
worldwide

Contact partners
Jürgen Thielert (sales director)

Further information
Development, manufacturing and sales of low-pressure mixing and dosing systems as well as automation concepts for semi and fully automated systems or stand-alone solutions for the processing and dosing of 1-component and 2-component polymer sealing, potting and gluing systems.

Range of Products

Types of adhesives
2-Component reaction adhesives based on PU

Types of sealants
2C PU, 2C Silicone, 1C PVC

Systems/procedures/Accessories/Services
Application systems (1C systems, 2C/multi-component systems, robots), automation for the mixing and dosing technology for adhesive applications, services, contract manufacturing

For applications in the field of
Automotive, packaging, electronics, aviation-, filter- and lighting industry, machine and device construction, household (white goods), photovoltaics, solar thermal energy, air condition, switch board enclosures.

Sulzer Mixpac Ltd.
Ruetistraße 7
CH-9469 Haag
Phone +41 (0) 81 772 20 00
Fax +41 (0) 81 772 20 01
Email: mixpac@sulzer.com
www.sulzer.com

Company

Year of formation
Sulzer Mixpac is a merger of successful, former independent companies with already existing business connections and coopera-tions. The formal acquisition was realized 2007.

Size of workforce
850 employees worldwide

Distribution
Direct distribution to producers of adhesives. toll filler and distribution partners for official trade

Contact partners
Switzerland:
Sulzer Mixpac Ltd.
Email: mixpac@sulzer.com

China:
Sulzer Mixpac China
Email: mixpac@sulzer.com

Untited Kingdom:
Sulzer Mixpac UK
Email: mixpac@sulzer.com

United States:
Sulzer Mixpac USA Inc.
Email: mixpac@sulzer.com

Range of Products

Manufacturer and supplier of metering, mixing, coatings and dispensing systems for reactive multi component material, offering comprehensive systems for various cartridge based 2-K applications.

Cartridge based 2-K systems with different volumes between 2.5 ml and 1.500 ml, with mixing rations of 1 : 1, 2 : 1, 3 : 1, 4 : 1 and 10 : 1
Manual, pneumatic and electric dispensers for 1K and 2K adhesives and sealant applications.
Full range of mixers for cartridge and dispensing machine applications
Brands: MIXPAC™, QUADRO™, MK™, COX™, MIXCOAT™, Statomix

Local Dispensing Solutions Worldwide

TechconSystems
Eagle Close, Chandlers Ford
Hampshire SO53 4NF
Phone +49 (0) 3222 109 1900
Fax + 44 (0) 2380 489 109
Email: Europe@techconsystems.com
www.techconsystems.com

Company

Year of formation
1961

Size of workforce
225

Ownership structure
Techcon Systems is a brand of OK International and owned by Dover

Sales channels
Techcon Systems sells through dedicated distributors and resellers in the local countries as well as direct to individual high volume consumers, such as formulators.

Contact partners
Sales Director Europe:
Rick Nutall
Email: rnutall@okinternational.com

Application Technology and Sales:
Packaging Materials Manager
Laurens Koch
Phone +49 (0) 172 618 52 76
Email: lkoch@techconsystems.com

Regional Sales Manager Nothern- and Western Europe:
André Tailleur
Phone + 44 (0) 7557 767 498
Email: atailleur@techconsystems.com

Regional Sales Manager Southern- and Eastern Europe:
Domenico Carluccio
Phone + 49 (0) 172 61 85 212
Email: dcarluccio@techconsytems.com

Technical Support Engineer
Ian Jennings
Phone + 44 (0) 23 80 48 90 05
Email: ijennings@techconsystems.com

Range of Products

Equipment, plant and components
Dispenser and application systems for 1 and 2 component materials, dispensing valves, Components for mixing, metering and dispensing technology

For applications in the field of
Electronics
Mechanical engineering and equipment construction
Automotive industry, aviation industry

Further information
Whether your goal is cost savings or process improvement, we can tailor a solution for your application.

Techcon Systems portfolio includes:

Bench Top:
Table Top Robots, Dispensers and controllers, disposable syringes from 3cc to 55cc capacities, premium dispensing tips, including straight, tapered, metal tapered, Teflon® lined, UV-block, full metal, flexible and bent versions in a variety of diameters and lengths

Cartridges:
Disposable plastic cartridges and components, 65 ml up to 595 ml (2.5 oz to 20 oz), wide range of disposable dispensing nozzles including bent versions

Techkit Systems - high end 2-component disposable dynamic mixing cartridges, Techkit Mixer (Automatic mixing of cartridges), high end industrial dispensing guns

High End Dispensing Valves:
Jet valves, Rotary Valves, Diaphragm Valves, Spool Valves, Pinch Tube Valves, Spray Valves, Hot Melt Valves

Other:
Application testing, pressure tanks, comprehensive fluid adaptor offering, spatulas and scrapers

t-s-i.de Misch- und Dosiertechnik GmbH
Bitscher Straße 6
D-66957 Vinningen
Phone +49 (0) 6335-9164-0
Fax +49 (0) 6335-9164-20
Email: info@t-s-i.de
www.t-s-i.de

Company

Year of formation
1998

Size of workforce
50 employees

Nominal capital
75 000 €

Ownership structure
GmbH

Subsidiaries
t-s-i.ro
t-s-i.pl
t-s-i.ae

Contact partners
Management:
Vanessa Schwartz

Anwendungstechnik und Vertrieb:
Thomas Schwartz

Further information
At t-s-i.de Misch-und Dosiertechnik GmbH, we are dedicated to a key mission: absolute reliability. We provide clients with innovative plant technology for processing sealants and adhesives.
Our services range from the engineering of standard plants to the creation of tailored end-to-end solutions and the use of robotics. They also include consulting, maintenance, and spare parts services. Comprehensive and capable client service is a top priority for us, and key to our

Range of Products

Equipment, plant and components
for conveying, mixing, metering and for adhesive application

For applications in the field of
Wood/furniture industry
Construction industry, including floors, walls and ceilings
Electronics
Mechanical engineering and equipment construction
Automotive industry, aviation industry
Household, recreation and office

concept of client relationships based on reliability. Whenever you are in need of bonding, mixing, and metering expertise and technologies, you can trust us to be a dependable partner. No matter whether your business is automotive, insulating glass, window bonding, renewable energies, or something completely else, we will provide you with solutions that work.

ViscoTec Pumpen- u. Dosiertechnik GmbH
Amperstraße 13
84513 Töging a. Inn
Phone +49 (0) 8631 9274-0
Fax +49 (0) 8631 9274-0
Email: mail@viscotec.de
www.viscotec.de

Company

Year of formation
1997

Size of workforce
150 employees

Managing partners
Dipl.-Ing. Georg Senftl,
Dipl.-Ing. Martin Stadler

Subsidiaries
• Töging am Inn, Germany (Headquarter)
• Georgia, USA
• Singapore
• Shanghai, China

Sales channels
worldwide

Contact partners
Management:
Christian Heidinger,
Senior Manager Business Unit Adhesives & Chemicals

Application technology and sales:
Daniel Pössnicker, International Sales Manager Business Unit Adhesives & Chemicals

Further information
ViscoTec Pumpen- u. Dosiertechnik GmbH primarily deals with systems required for conveying, dosing, applying, filling and empty-ing fluids ranging from medium to high viscosity.

Range of Products

Equipment, plant and components
Systems for conveying, dosing, applying, filling and emptying fluids ranging from medium to high viscosity:
• Dispenser and dosing systems (applying and administering, 1K-/2K-dosing, potting, filling, process dosing, spraying)
• Emptying and supplying systems
• Preparing systems
• Complete systems
• Dedicated equipment

For applications in the field of
• Automotive
• Aerospace
• Electronics
• General industry
• Renewable energies
• Plastics
• Food
• Cosmetics
• Pharmaceuticals
• Medical technology
• Biotechnology
• 3D printing

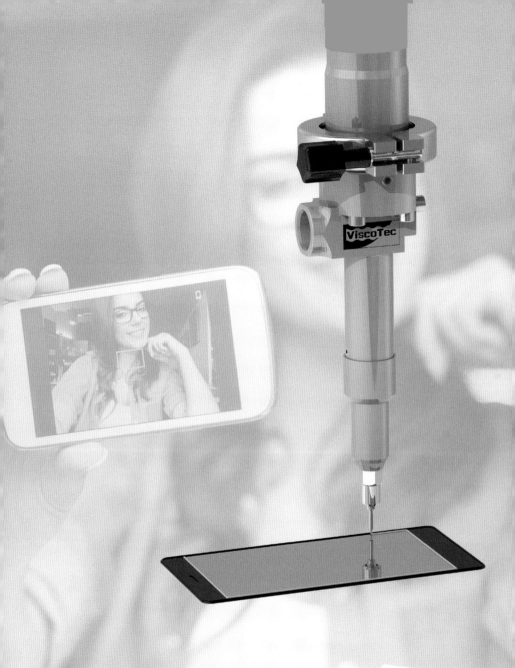

Walther Spritz- und Lackiersysteme GmbH
a Member of WAGNER GROUP
Kärntner Straße 18 - 30
D-42327 Wuppertal
Phone +49 (0) 2 02-7870
Fax +49 (0) 2 02-7 87-22 17
Email: info@walther-pilot.de
www.walther-pilot.de

Company

Size of workforce
142 employees

Locations
Wuppertal-Vohwinkel
Neunkirchen-Struthütten

Managing partners
Wilhelm W. Schmidts

Contact partners
Sales Manager:
Erik Niehaus

Applications technology:
Torsten Bröker
Benno Burggräf

Distribution channels
Field sales staff in Germany and
representatives in Europe and overseas.

Range of Products

Systems and components for the application
of adhesives, sealants and paints. Functioning
as a systems supplier, WALTHER PILOT
engineers customized, all-round solutions.
They guarantee the best results over the
long run in terms of economy, user
friendliness and environment protection.

Application
Adhesive spray guns and automated
applicators
Extrusion guns
Metering valves
Spray guns for dots and lines
Multi-component metering and
mixing systems

Material conveyance
Pressure tanks
Diaphragm pumps
Piston pumps
Pump systems for high-viscosity media
Supply stations
Systems for shear-sensitive materials

Overspray exhaust
Spray booths
Filter technology
Ventilation systems

COMPANY PROFILES

Adhesive Technology
Consultancy Companies

ChemQuest Europe Inc.

Bilker Straße 27
D-40213 Düsseldorf
Phone +49 (0) 211-4 36 93 79
Fax +49 (0) 32 12-1 07 16 75
Mobile: +49 (0)1 71-3 41 38 38
Email: jwegner@chemquest.com
www.chemquest.com

Member of IVK

Company

Contact partners
Dr. Jürgen Wegner,
Managing Director
Email: jwegner@chemquest.com

Dr. Hubertus von Voithenberg,
Managing Director
Email: hvoithenberg@chemquest.com

CV and professional background
under www.chemquest.com

Consultancy

The CemQuest Group is an international consulting firm headquartering in Cincinnati, Ohio, USA, with branch offices in Europe, China, North Africa and Latin America. We specialize in consulting the Adhesives, Sealants, Construction Chemicals and Coatings Industry through all steps within the value chain from raw materials manufacturing through product formulation to all types of industrial and non-industrial end use applications.

Based on in-depth knowledge our service portfolio includes all types of Management Consulting, M&A activities, market research plus market and technology trend analysis. All associates of ChemQuest are experienced professionals from within the Adhesives, Sealants or Coatings Industry, and we are partnering with IFAM Fraunhofer Bremen in certified Adhesives education in North America. For further information and contact details please visit our website www.chemquest.com

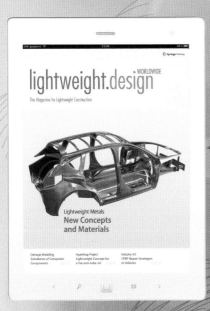

Our know-how – your future!

HINTERWALDNER CONSULTING

Consulting Chemists & Business Economists since 1956

Hinterwaldner Consulting
Dipl.-Kfm. Stephan Hinterwaldner
Markplatz 9
D-85614 Kirchseeon
Phone +49 (0) 80 91-53 99-0
Fax +49 (0) 80 91-53 99-20
Email: info@HiwaConsul.de
www.HiwaConsul.de

Member of IVK

Company

Year of formation
1956

Ownership structure
Privately held

Contact partner
Dipl.-Kfm. Stephan Hinterwaldner

Adhesive Consulting
info@HiwaConsul.de

Adhesive Conferencing
contact@mkvs.de
contact@in-adhesives.com

Consultancy

Global Expert Research and Technology Consulting and Conferencing in the World of Adhesives

- Raw Materials, Ingredients, Intermediates, Additives
- Formulations, Applications, Product and Process Developments, Feasibility Studies
- Adhesives, Adhesive Tapes, Coatings, Cementations, Glues, Sealants, Release Liners
- Pressure Sensitive Adhesives, Hot Melt Adhesives, Chemically and Radiation Curing Adhesives Systems, Structural Bonding, Structural Glazing Systems
- Technologies in Adhesive Bonding, Coating, Converting, Film, Foil, Labeling, Laminating, Lightweight, Metalizing, Packaging, Printing, Sealing, Surface
- Polymer, Chemically and Radiation Curing, Petro-based, Bio-based and Green Chemistries
- Cosmetics and Toiletries, Beauty and Personal Care, Home Care
- Natural, Renewable, Sustainable, Bio-based and Certified Organic Products

Organizer/Co-organizer
- Munich Adhesives and Finishing Symposium www.mkvs.de
- in-adhesives Symposium www.in-adhesives.com
- European Coatings Congress

Klebtechnik Dr. Hartwig Lohse e. K.

Hofberg 4
D-25597 Breitenberg
Phone +49 (0) 48 22-9 51 80
Fax +49 (0) 48 22-9 51 81
Email: hl@hdyg.de
www.how-do-you-glue.de

Member of the IVK

Company

Year of formation
2009

Contact partners
Dr. Hartwig Lohse

Further information
Based on 30+ years of industrial experience in the development, technical service
and technical marketing of adhesives we
are providing competent consultancy
services along the adhesive supply chain.
Our customers are industrial users of
adhesives, adhesives manufacturing and
raw material supply companies, equipment
and plant manufacturing companies as well
as manufacturers of dispensing and testing
equipment.

Consultancy

Our consultancy service portfolio includes:

- optimisation of current and the planning
  of new bonding processes incl. generation of the specific requirements profile,
  selection of the best adhesive technology,
  source search for commercial adhesives,
  evaluation of adhesives, bond design
  advice, selection of suitable surface pretreatment measures, selection of suitable
  dispensing technology, advice on quality
  assurance measures, EH&S, ...
- implementation of DIN 2304 (Adhesive
  bonding technology - Quality requirements for adhesive bonding processes)
- adhesive failure analysis (troubleshooting)
- adhesive education, training at all levels
  specific to the clients requirements
- development of adhesives as well as raw
  materials for the use in manufacturing of
  adhesives
- market research, market & market trend
  analysis
- ...

For further information please visit us at
www-how-do-you-glue.de

Contract Manufacturing
and Filling Services

chemisch technische Produkte

Am Nordturm 5
D-46562 Voerde
Phone +49 (0) 2 81-8 31 35
Fax +49 (0) 2 81-8 31 37
Email: mail@loop-gmbh.de

Member of IVK

Company

Year of formation
1993

Size of workforce
25

Contact partners
Managing Director:
DI Marc Zick

Production Manager:
Jürgen Stockmann

Further information
LOOP is consistently expanding its personnel and equipment at plant and the KST Customer Service Test Centre.

LOOP has a silicone handling division at a separate location.

LOOP is partner of international and national well known companies.

Range of Products

Toll-Production of the following product groups
- polymer formulations, one- and two- component systems (unfilled, filled), aqueous, solvent based and solvent-free
- additives/concentrates of actives substances (liquid, pasty, powdery)
- impregnating and mould resin systems
- slurries
- powder mixtures
- substrates like SiO_2 coated with active substances
- Concentrate pastes
- Granulates and their fractions for applications in engineering and decorative applications
- compounds for electrical components and diverse additional product groups

LOOP operates on the following sectors for its partners
- adhesives and sealants
- polymer and binder chemicals
- pigments and fillers
- additives
- construction chemicals/preservation of buildings
- composites
- foundry products
- electronics
- cables
- power generation (wind power, photovoltaics)
- glass
- textile and paper finishes
- wood preservatives
- varnishes, paints, and printing inks
- professional development, laboratory, and consultancy services
- large scale distribution, etc.

Research and Development

Fraunhofer

IFAM

Fraunhofer Institute for Manufacturing
Technology and Advanced Materials IFAM
– Adhesive Bonding Technology
and Surfaces –

Wiener Straße 12
D-28359 Bremen
Phone +49 (0) 4 21-22 46-4 00
Fax +49 (0) 4 21-22 46-4 30
Email: info@ifam.fraunhofer.de
www.ifam.fraunhofer.de

Member of IVK

Company

Year of formation
1968

Size of workforce
All in all 609
Division of Adhesive Bonding Technology
and Surfaces: > 300

Ownership structure
The Fraunhofer IFAM is a constituent
entity of the Fraunhofer-Gesellschaft zur
Förderung der angewandten Forschung e. V.
and as such has no separate legal status

Contact
**Fraunhofer IFAM
– Adhesive Bonding Technology and
Surfaces –**
Director: Prof. Dr. Bernd Mayer
Deputy director:
Prof. Dr. Andreas Hartwig

Work Areas

R&D – Contract-research and development
– in all fields of adhesive bonding technology and surfaces as well as materials, in addition providing certifying training courses
in adhesive bonding technology and fiber
composite materials:

Adhesives and Polymer Chemistry:
Prof. Dr. Andreas Hartwig
Phone +49 (0) 4 21-22 46-4 70
Email: andreas.hartwig@ifam.fraunhofer.de

Adhesive Bonding Technology:
Dr. Holger Fricke
Phone +49 (0) 4 21-22 46-6 37
Email: holger.fricke@ifam.fraunhofer.de

Work Areas

Material Science and Mechanical Engineering:
Dr. Markus Brede
Phone +49 (0) 4 21-22 46-4 76
Email: markus.brede@ifam.fraunhofer.de

*Automation and Production Technology
(CFK Nord, Stade):*
Dr. Dirk Niermann
Phone +49 (0) 41 41-7 87 07-1 01
Email: dirk.niermann@ifam.fraunhofer.de

Adhesion and Interface Research:
Dr. Stefan Dieckhoff
Phone +49 (0) 4 21-22 46-4 69
Email: stefan.dieckhoff@ifam.fraunhofer.de

Plasma Technology and Surfaces PLATO:
Dr. Ralph Wilken
Phone +49 (0) 4 21-22 46-4 48
Email: ralph.wilken@ifam.fraunhofer.de

Paint/Lacquer Technology:
Dr. Volkmar Stenzel
Phone +49 (0) 4 21-22 46-4 07
Email: volkmar.stenzel@ifam.fraunhofer.de

*Workforce Qualification and Technology
Transfer*
*– Training Center for Adhesive Bonding
Technology:* Dr. Erik Meiß
Phone +49 (0) 4 21-22 46-6 32
Email: erik.meiss@ifam.fraunhofer.de
www.bremen-bonding.com

*– Training Center for Fiber Composite
Technology:* Beate Brede
Phone +49 (0) 4 21-56 65-4 65
Email: beate.brede@ifam.fraunhofer.de
www.bremen-composites.com

ZHAW School of Engineering

Laboratory of Adhesives and Polymer Materials
Institute of Materials and Process Engineering (IMPE)
Technikumstrasse 9, CH-8401 Winterthur
Phone +41 (0) 58 934 65 86
Email: christof.braendli@zhaw.ch
www.zhaw.ch/impe

Member of FKS

Company

Year of formation
School of Engineering: 1874, Institute: 2007

Size of workforce
School of Engineering: 580, Institute: 40

Managing partners
School of Engineering: Prof. Dr. Martina Hirayama, Phone: +41 58 934 73 26, Email: martina.hirayama@zhaw.ch

Institute of Materials and Process Engineering: Prof. Dr. Andreas Amrein, Phone: +41 58 934 73 51, Email: andreas.amrein@zhaw.ch

Laboratory of Adhesives and Polymer Materials. Prof. Dr. Christof Brändli, Phone +41 58 934 65 86, Email: christof.braendli@zhaw.ch

Ownership structure
Part of the Zurich University of Applied Sciences (ZHAW)

Contact partners
Laboratory of Adhesives and Polymer Materials: Prof. Dr. Christof Brändli, Phone +41 58 934 65 86, Email: christof.braendli@zhaw.ch

Further information
Applied R&D in the fields of materials, adhesive bonding, and surfaces. Adhesive development and testing. Broad range of adhesive competencies including adhesive chemistry and analysis.

Work Areas

Types of adhesives
Hot melt adhesives
Reactive adhesives
Solvent-based adhesives
Dispersion adhesives
Pressure-sensitive adhesives

Synthesis and Formulation
Adhesive formulation and synthesis
Polymer compounding, reactive extrusion, and synthesis
Online reaction control with IR spectroscopy
Functionalization of nano particles

Characterization
Adhesive performance tests
Curing behavior studies
Thermal and mechanical analysis
Flow properties determination with rheological methods
Morphological and surface analysis

Applications
Adhesive development
(structure-property relationship)
Polymer development
(incl. tapes, film, grafting, reactive, ...)
Nanomodifications

INSTITUTES AND RESEARCH FACILITIES

Research and Development

Adhesive bonding technology is a key contributor to the development of innovative products, creating the conditions for new, future-proof markets to open up across all industries. Small and medium-sized companies benefit from new joining methods by developing sophisticated products which give them competitive edge.

In order to profit from the advantages that adhesive bonding technology offers over other methods of joining, it is important to include the whole process from product planning through quality management to staff training.

This can only be achieved by close cooperation between research institutions and industry so that research results are applied quickly to the development of innovative products and production processes.

The following list includes all known research facilities and institutes which are committed to working with their partners in industry to resolve adhesive issues across a range of areas.

Deutsches Institut für Bautechnik
(Federal Agency for the Approval of
Construction Products)
Kolonnenstraße 30 B
D-10829 Berlin

Contact:
Dr. Patricia Döring
Phone: +49 (0) 30 78730 220
Fax: +49 (0) 30 78730 11220
Email: pdo@dibt.de
www.dibt.de

FH Aachen - University of Applied Sciences
Füge- und Trenntechnik
(Joining and Separation Technology)
Goethestraße 1
D-52064 Aachen

Contact:
Prof. Dr.-Ing. Markus Schleser
Phone: +49 (0) 241 6009 52385
Fax: +49 (0) 241 6009 52368
Email: schleser@fh-aachen.de
www.fh-aachen.de

Fogra Forschungsgesellschaft Druck e.V.
(Graphic Technology Research Association)
Streitfeldstraße 19
D-81673 München

Contact:
Dr. Eduard Neufeld
Phone: +49 (0) 89 43182 112
Fax: +49 (0) 89 43182 100
Email: info@fogra.org
www.fogra.org

FOSTA Forschungsvereinigung
Stahlanwendung e.V.
(Research Association for Steel Application)
Stahl-Zentrum
Sohnstraße 65
D-40237 Düsseldorf

Contact:
Dr.-Ing. Hans-Joachim Wieland
Phone: +49 (0) 211 6707 426
Fax: +49 (0) 211 6707 840
Email: fosta@stahlforschung.de
www.stahl-online.de

Fraunhofer-Institut für Fertigungstechnik und
Angewandte Materialforschung – IFAM
(Fraunhofer Institute for Manufacturing
Technology and Advanced Materials – IFAM)
Wiener Straße 12
D-28359 Bremen

Contact:
Prof. Dr. Bernd Mayer
Phone: +49 (0) 421 2246 419
Fax: +49 (0) 421 2246 774401
Email: bernd.mayer@ifam.fraunhofer.de
Prof. Dr. Andreas Groß
Phone: +49 (0) 421 2246 437
Fax: +49 (0) 421 2246 605
Email: andreas.gross@ifam.fraunhofer.de
www.ifam.fraunhofer.de

Fraunhofer-Institut für Holzforschung –
Wilhelm-Klauditz-Institut – WKI
(Fraunhofer Institute for Wood Research –
Wilhelm-Klauditz-Institute (WKI))
Bienroder Weg 54 E
D-38108 Braunschweig

Contact:
Dr. Heike Pecher
Phone: +49 (0)531 2155 206
Email: heike.pecher@wki.fraunhofer.de
www.wki.fraunhofer.de

Fraunhofer-Institut für Werkstoff- und
Strahltechnik – IWS
(Fraunhofer Institute Material and Beam
Technology – IWS)
(Klebtechnikum an der TU Dresden, Institut für
Fertigungstechnik,Professur für Laser- und
Oberflächentechnik)
Winterbergstraße 28
D-01277 Dresden

Contact:
Dr. Irene Jansen
Phone: +49 (0) 351 463 35210
Fax: +49 (0) 351 463 37755
Email: irene.jansen@iws.fraunhofer.de
www.iws.fraunhofer.de

Fraunhofer-Institut für Zerstörungsfreie
Prüfverfahren – IZFP
(Fraunhofer Institute for Nondestructive
Testing – IZFP)
Campus E3.1
D-66123 Saarbrücken

Contact:
Prof. Dr.-Ing. Bernd Valeske
Phone: +49 (0) 681 9302 3610
Fax: +49 (0) 681 9302 11 3610
Email: bernd.valeske@izfp.fraunhofer.de
www.izfp.fraunhofer.de

Johann Heinrich von Thünen Institut (vTI)
Bundesforschungsistitut für Ländliche Räume,
Wald und Fischerei
Institut für Holzforschung
(Johann Heinrich von Thünen Institute (vTI)
Institute of Wood Research)
Leuschnerstraße 91
D-21031 Hamburg

Contact:
Dr. Dr. h. c. Uwe Schmitt
Phone: +49 (0) 40 73962 400
Fax: +49 (0) 40 73962 499
Email: uwe.schmitt@thuenen.de
www.thuenen.de

Hochschule München
Institut für Verfahrenstechnik Papier e.V. (IVP)
(Institute of Munich University of
Applied Science)
Schlederloh 15
D-82057 Icking

Contact:
Prof. Dr. Stephan Kleemann
Phone: +49 (0) 89 1265 1668
Fax: +49 (0) 89 1265 1560
Email: kleemann@hm.edu
www.hm.edu

Hochschule für nachhaltige Entwicklung
Eberswalde (FH)
Fachbereich Holzingenieurwesen
(Eberswalde University for Sustainable
Development
Department of Wood Engineering)
Schlickerstraße 5
D-16225 Eberswalde

Contact:
Prof. Dr.-Ing. Ulrich Schwarz
Phone: +49 (0) 3334 657 371
Fax: +49 (0) 3334 657 372
Email: ulrich.schwarz@hnee.de
www.hnee.de/holzingenieurwesen

IFF GmbH
Induktion, Fügetechnik, Fertigungstechnik
(Induction and Joining Technology Engineering)
Gutenbergstraße 6
D-85737 Ismaning

Contact:
Prof. Dr.-Ing. Christian Lammel
Phone: +49 (0) 89 9699 890
Fax: +49 (0) 89 9699 8929
Email: christian.lammel@iff-gmbh.de
www.iff-gmbh.de

ift Rosenheim GmbH
Institut für Fenstertechnik e.V.
(Institute for Window Technology)
Theodor-Gietl-Straße 7-9
D-83026 Rosenheim

Contact:
Prof. Ulrich Sieberath
Phone: +49 (0) 80 31261 0
Fax: +49 (0) 80 31261 290
Email: info@ift-rosenheim.de
www.ift-rosenheim.de

ihd – Institut für Holztechnologie Dresden GmbH
(Institute of Wood Technology)
Zellescher Weg 24
D-01217 Dresden

Contact:
Dr. rer. nat. Steffen Tobisch
Phone: +49 (0) 351 4662 257
Fax: +49 (0) 351 4662 211
Mobil: +49 (0) 1622 696330
Email: steffen.tobisch@ihd-dresden.de
www.ihd-dresden.de

Institut für Holzbiologie und Holztechnologie
(Institute for Wood Biology and Wood
Technology)
Büsgenweg 4
D-37077 Göttingen

Contact:
Prof. Dr. Holger Militz
Phone: +49 (0) 551 393541
Fax: +49 (0) 551 399646
Email: hmilitz@gwdg.de
www.uni-goettingen.de

Institut für Oberflächen- und Fertigungstechnik
Professur Fügetechnik und Montage
TU Dresden
(Institute of Manufacturing Technology, Joining
Technology and Assembly
Technical University of Dresden)
George-Bähr-Straße 3c
D-01069 Dresden

Contact:
Prof. Dr.-Ing. habil. Uwe Füssel
Phone: +49 (0) 351 46337 615
Fax: +49 (0) 351 46337 249
Email: uwe.fuessel@tu-dresden.de
https://tu-dresden.de/ing/
maschinenwesen/if/fue

IVLV - Industrievereinigung für Lebens-
mitteltechnologie und Verpackung e.V.
(Industry Association for Food Technology and
Packaging)
Giggenhauser Straße 35
D-85354 Freising

Contact:
Dr.-Ing. Tobias Voigt
Phone: +49 (0) 8161 491140
Fax: +49 (0) 8161 491142
Email: tobias.voigt@ivlv.org
www.ivlv.org

iwb – Anwenderzentrum Augsburg
Technische Universität München
(iwb – Augsburg Application Centre
Technical University of Munich)
Beim Glaspalast 5
D-86153 Augsburg

Contact:
Dipl.-Ing. Johannes Glasschröder
Phone: +49 (0) 821 56883 53
Fax: +49 (0) 821 56883 50
Email: johannes.glasschroeder@
iwb.mw.tum.de
www.iwb.mw.tum.de

Kompetenzzentrum Werkstoffe der Mikrotechnik
Universität Ulm
(WMtech University of Ulm)
Albert-Einstein-Allee 47
D-89081 Ulm

Contact:
Prof. Dr. Hans-Jörg Fecht
Phone: +49 (0) 731 50254 91
Fax: +49 (0) 731 50254 88
Email: info@wmtech.de
www.wmtech.de

Leibniz-Institut für Polymerforschung
Dresden e.V.
(Leibniz Institute of Polymer Research Dresden)
Hohe Straße 6
D-01069 Dresden

Contact:
Prof. Dr. Edith Mäder
Phone: +49 (0) 351 46583 05
Fax: +49 (0) 351 46583 62
Email: emaeder@ipfdd.de
www.ipfdd.de

Naturwissenschaftliches und Medizinisches
Institut an der Universität Tübingen
(The Natural and Medical Sciences
Institute at the University of Tübingen)
Markwiesenstraße 55
D-72770 Reutlingen

Contact:
Sebastian Wagner
Phone: +49 (0)7121 51530 523
Fax: +49 (0)7121 51530 62
Email: sebastian.wagner@nmi.de
www.nmi.de

ofi Österreichisches Forschungsinstitut
für Chemie und Technik
Institut für Klebetechnik
(ofi Austrian Research Institute for Chemistry
and Technology
Institute for Adhesive Technology)
Viktor-Kaplan-Straße 2 / Bauteil E
A-2700 Wiener Neustadt

Contact:
Ing. Michael Bodendorfer
Phone: +43 (0)1798 1601 670
Fax: +43 (0)1798 1601 303
Email: michael.bodendorfer@ofi.at
www.ofi.at

Papiertechnische Stiftung PTS
(Paper Technology Foundation)
Heßstraße 134
D-80797 München

Contact:
Dr. rer. nat. Frank Miletzky
Phone: +49 (0) 89 12146 184
Fax: +49 (0) 89 12146 36
Email: frank.miletzky@ptspaper.de
www.ptspaper.de

Prüf- und Forschungsinstitut Pirmasens e.V.
(Test and Research Institute for Footwear
Production)
Marie-Curie-Straße 19
D-66953 Pirmasens

Contact:
Dr. Kerstin Schulte
Phone: +49 (0)6331 2490 0
Fax: +49 (0)6331 2490 60
Email: info@pfi-germany.de
www.pfi-pirmasens.de

RWTH Aachen
ISF - Institut für Schweißtechnik und
Fügetechnik
(RWTH Aachen
ISF – Welding and Joining Institute)
Pontstraße 49
D-52062 Aachen

Contact:
Prof. Dr. -Ing. Uwe Reisgen
Phone: +49 (0) 241 80 93870
Fax: +49 (0) 241 80 92170
Email: office@isf.rwth-aachen.de
www.isf.rwth-aachen.de

Technische Universität Berlin
Fügetechnik und Beschichtungstechnik im
Institut für Werkzeugmaschinen und
Fabrikbetrieb
(Technical University Berlin
Joining and Coating Technology Departments
of Machine Tools and Factory Management)
Pascalstraße 8 – 9
D-10587 Berlin

Contact:
Prof. Dr.-Ing. habil. Christian Rupprecht
Phone: +49 (0) 30 314 25176
Email: info@fbt.tu-berlin.de
www.fbt.tu-berlin.de

Technische Universität Braunschweig
Institut für Füge- und Schweißtechnik
(Technical University Braunschweig
Institute of Joining and Welding Technology)
Langer Kamp 8
D-38106 Braunschweig

Contact:
Univ.-Prof. Dr.-Ing. Prof. h.c. Klaus Dilger
Phone: +49 (0) 531 391 95500
Fax: +49 (0) 531 391 95599
Email: k.dilger@tu-braunschweig.de
www.ifs.ing.tu-bs.de

Technische Universität Kaiserslautern
Fachbereich Maschinenbau und
Verfahrenstechnik
Arbeitsgruppe Werkstoff- und Oberflächen-
technik Kaiserslautern (AWOK)
(Technical University Kaiserslautern
Work Group for Materials and Surface
Technologies)
Gebäude 58, Raum 462
Erwin-Schrödinger-Straße
D-67663 Kaiserslautern

Contact:
Univ.-Prof. Dr.-Ing. Paul Ludwig Geiß
Phone: +49 (0) 631 205 4117
Fax: +49 (0) 631 205 3908
Email: geiss@mv.uni-kl.de
www.mv.uni-kl.de/awok

TechnologieCentrum Kleben
TC-Kleben GmbH
(Center Adhesive Bonding Technology)
Carlstraße 50
D-52531 Übach-Palenberg

Contact:
Dipl.-Ing. Julian Brand
Phone: +49 (0) 2451 9712 00
Fax: +49 (0) 2451 9712 10
Email: post@tc-kleben.de
www.tc-kleben.de

Universität Kassel
Institut für Werkstofftechnik, Kunststofffüge-
techniken, Werkstoffverbunde
(University of Kassel
Institute für Materials Engineering)
Mönchebergstraße 3
D-34125 Kassel

Contact:
Prof. Dr.-Ing. H.-P. Heim
Phone: +49 (0) 561 80436 70
Fax: +49 (0) 561 80436 72
Email: heim@uni-kassel.de
www.uni-kassel.de/maschinenbau

Universität Kassel
Fachgebiet Trennende und Fügende Fertigung-
verfahren
(University of Kassel
Department of Separative and Joining
Production Processes)
Kurt-Wolters-Straße 3
D-34125 Kassel

Contact:
Prof. Dr.-Ing. Prof. h.c. Stefan Böhm
Phone: +49 (0) 561804 3141
Fax: +49 (0) 561804 2045
Email: s.boehm@uni-kassel.de
www.tff-kassel.de

Universität des Saarlandes
Adhäsion und Interphasen in Polymeren
(University of Saarland
Adhesion and Interphases in Polymers)
Campus, Geb. C6.3
D-66123 Saarbrücken

Contact:
Prof. rer. nat. Wulff Possart
Phone: +49 (0) 681 302 3761
Fax: +49 (0) 681 302 4960
Email: w.possart@mx.uni-saarland.de
www.uni-saarland.de/nc/startseite

Universität Paderborn
Laboratorium für Werkstoff- und Fügetechnik
(University of Paderborn
Laboratory for Materials and Joining Technology)
Pohlweg 47-49
D-33098 Paderborn

Contact:
Prof. Dr.-Ing. Gerson Meschut
Phone: +49 (0) 5251 603031
Fax: +49 (0) 5251 603239
Email: meschut@lwf.upb.de
www.lwf-paderborn.de

Wehrwissenschaftliches Institut für Werk- und
Betriebsstoffe - WIWeB
(Bundeswehr Research Institute for Materials,
Explosives, Fuels and Lubricants – WIWeB)
Institutsweg 1
D-85435 Erding

Contact:
Dr. Jürgen von Czarnecki
Phone: +49 (0) 81 229590 0
Fax: +49 (0) 81 229590 3902
Email: wiweb@bundeswehr.org
http://www.baainbw.de/portal/a/baain/
start/diensts/wiweb

Westfälische Hochschule Abteilung
Recklinghausen
Fachbereich Wirtschaftsingenieurwesen
Organische Chemie und Polymere
(Westfälische Hochschule
Department of Industrial Engineering,
Organic Chemistry and Polymers)
August-Schmidt-Ring 10
D-45665 Recklinghausen

Contact:
Prof. Dr. Klaus-Uwe Koch
Phone: +49 (0) 2361 915 456
Fax: +49 (0) 2361 915 751
Email: klaus-uwe.koch@w-hs.de
www.w-hs.de/erkunden/fachbereiche/
wirtschaftsingenieurwesen/portrait-des-
fachbereichs/

German
Adhesives
Association
Industrieverband Klebstoffe e.V.

ANNUAL REPORT 2016

Economic Report

The German adhesives industry continued to grow in 2015 and 2016. During 2015, the industry saw a nominal increase in domestic revenues of 2.2 percent across all types of adhesive products, in other words, adhesives, sealants, cement-based construction materials and adhesive tapes. In addition, innovative technologies and exchange rate effects boosted the export business, with revenues growing by 3 percent. In 2016, the German adhesives industry achieved a 1.2 percent nominal increase in domestic revenues to € 3.75 billion, while exports remained stable. In 2017, the industry expects continued domestic growth of up to 2.0 percent. The positive economic situation of the construction industry is a significant driver of growth in this area.

The German adhesives industry is in a very strong position on European and other international markets. With a global market share of around 19 percent, it is the world market leader. It also holds the top positions on the western European market, with adhesives consumption of 28 percent and a production share of more than 36 percent.

Worldwide, sales revenues of approximately € 61 billion per year (not adjusted for exchange rate effects) were generated with adhesives, sealants, adhesive tapes and system products. The German adhesives industry, which consists mainly of medium-sized companies, plays a prominent role on the international stage. Most of the companies manufacture their products in Germany and export them worldwide. In addition, approximately 20 percent of them supply the world markets from their local adhesives factories outside Germany.

With both business models, the German adhesives industry generates sales revenues of almost € 11.3 billion worldwide. Exports from Germany amount to more than € 1.6 billion, while German adhesives manufacturers generate a further € 8.1 billion of sales revenues locally from their production facilities outside Germany. The German market has a sales volume of almost € 3.8 billion per year. Through the application of "Adhesive systems developed in Germany" in almost all manufacturing industries and in the construction sector, the adhesives industry was responsible for indirect value creation of significantly more than € 400 billion in Germany. Worldwide, value creation amounts to more than € 1 trillion.

This strong position is the direct result of innovative technological developments, for example in the fields of mechanical and plant engineering. In these sectors, the adhesives industry acts as a partner to supply practical, value-added solutions.

**The link between the price of crude oil and of basic raw materials
for adhesives weakens**
The price of crude oil can no longer be directly reflected in the price of end products. The connection has almost been broken. Prices are increasingly being determined by the many refining processes involved in the value chain (see the graphic "Raw material flows") and the availability of a raw material on the market. The link between the price of a range of basic raw materials and that of crude oil is likely to weaken even further. This is made very clear by a comparison between the fluctuations in the price of crude oil and of acetic acid, VAM, ethylene and methanol (see the graphic "Price fluctuations in crude oil and basic raw materials"). While

the price of crude oil fell throughout 2015, the prices of the raw materials referred to above remained largely stable or were heavily influenced by regional supply and demand. In particular in the case of high-quality adhesives, the fluctuations in prices are no longer linked because there are up to 10 processing stages between crude oil and the raw materials used to manufacture PUR, epoxy resin and acrylate adhesives.

About our Committee Work

General Assembly

The general assembly in 2016 in Berlin and in 2017 in Mainz took place against the background of a positive overall economic situation. Adhesives companies expect the market, exports, revenues and employee numbers to remain stable during the fiscal year 2017.

The German adhesives industry grew by 1.2% in 2016. The industry is expected to continue its positive economic development in 2017. Growth of at least 2% has been forecast for the domestic market and, from a global perspective, at least in certain areas the prognosis for exports is also good.

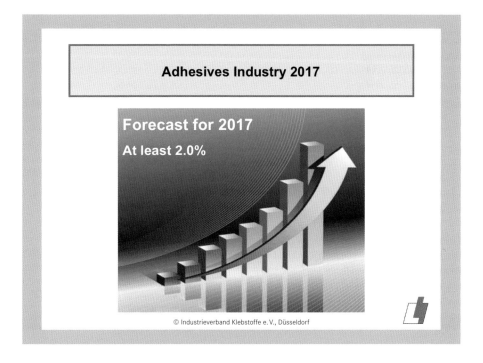

Adhesives Industry 2017

Forecast for 2017
At least 2.0%

© Industrieverband Klebstoffe e. V., Düsseldorf

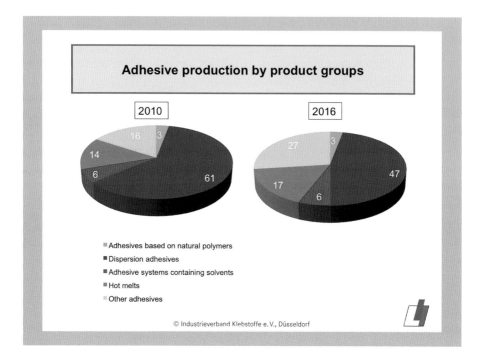

The services that the German Adhesives Association (IVK) provides for its members cover three areas:
- Technology
- Communication
- Future initiatives

IVK continues to give the highest priority to technical subjects because working together to manage and implement the legislative requirements in particular that apply to adhesives companies has a significant impact on the success of the industry. This applies to REACH, labelling (GHS/ CLP), permitted biocides, (mandated) standardisation and sustainability, as well as technical briefing notes, seminars and conferences and contacts with customer organisations.

The information document drawn up in cooperation with the German Environment Agency entitled "Hygienic assessment of anaerobic adhesives that come in contact with drinking water" was received positively by member companies.

After the agency withdrew its guidelines and positive lists in 2005, manufacturers of anaerobic thread sealants no longer had any means of obtaining the drinking water approvals (KTW certificates) required by the market. The approach taken by testing institutes at the time, which involved using testing methods that bore little resemblance to practical applications, coming to approximate conclusions about the influence of the cured adhesives on drinking water and then producing test certificates based on the guidelines that had been withdrawn, was highly unsatisfactory. Therefore, in 2013 IVK set up a working group consisting of the affected adhesives manufacturers. This group collaborated with the agency to establish a valid legal framework and develop a practical test which would allow for fair competition and ensure customer satisfaction.

The agency has now introduced the group of "marginal products" which includes anaerobic adhesives, primarily as a result of the close cooperation between the ad hoc IVK group and the agency. An important factor is that the raw materials used in these products do not need to be evaluated in this case, which means that there is no positive list for marginal products. In addition, there are no formulation-related requirements for individual materials or other additional requirements. However, the thread sealants must follow the outline formulation guidelines agreed with the agency.

The German Environment Agency has published the information document "Hygienic assessment of anaerobic adhesives that come in contact with drinking water" on its website: (http://www.umweltbundesamt.de/themen/wasser/trinkwasser/trinkwasser-verteilen). The agency does not consider a certificate of conformity to be necessary for anaerobic adhesives. The consistent quality of the formulation of anaerobic adhesives and the traceability of these products can be guaranteed by the manufacturers' own quality management systems. There is also the option of having an independent body verify that the formulation of an anaerobic adhesive corresponds with the formulation guidelines. This means that anaerobic thread sealants are once again covered by the provisions of legislation on drinking water, but with much less red tape and much lower costs.

It is well known that **sample EPDs** (sample environmental product declarations for construction adhesives) and guidelines are available on the IVK website in German and English at http://epd.klebstoffe.com/. These have now been certified by IBU (the German institute for construction and the environment) with the help of the consulting company Thinkstep on the basis of European data and published by FEICA (the Association of the European Adhesive & Sealant Industry) at http://www.feica.eu/our-priorities/key-projects/epds.aspx. They are widely recognised in Europe, but in some cases additional data is required. The EU Commission is working on developing EN 15804, the basic standard for EPDs, for the purposes of harmonisation.

In addition, IVK and its committees are actively working on a number of different issues and projects as part of the association's portfolio of services, which is based on the needs of its members. These include:
- Close involvement in the legislative procedure for the circular economy package on a national and a European level
- The introduction of the DIN 2304 standard "Quality requirements for adhesive bonding processes"
- Support by EMICODE for very low emission products as part of the IVK sustainability concept
- Publication of information on the approval of various in-can preservatives, the resulting changes in labelling requirements and the accompanying deadlines

A number of activities have been taking place in the field of communication. These include:

Der geklebte Glaspalast

- The magazine "Kleben fürs Leben" (Bonding for Life) which has an article on a variety of interesting subjects relating to all aspects of adhesives and bonding. The magazine is also available as an interactive e-paper and is published jointly by the adhesives industries of the German-speaking countries.
- Regular press releases have been issued that have received a response from a wide variety of media organizations. Many articles on the subject of bonding have been published in the print and online media, on the IVK press portal www.klebstoff-presse.com and on social media networks.
- A new promotional video has been launched about the German adhesives industry. The video is called "Faszination Kleben" (Fascinating World of Adhesives) and shows the wide-ranging applications and uses of adhesive systems. It is available on social networks.
- The Technical Committees have published a total of 47 technical briefing notes. Almost all of them are available in English. The technical briefing notes are an important and much valued source of information for the customers of the adhesives industry.
- Conferences and seminars have been held for members and their customers on a range of different technical and regulatory subjects. These events are also an ideal opportunity for networking, meeting existing contacts and making new ones.
- A history of the association was published for its 70th anniversary in 2016 entitled "Auf der Höhe der Zeit" (Keeping Pace with the Times). It describes the important stages in the association's development against the background of the economic, political and social history of the time. The active role that IVK has played in shaping a market characterised by technical progress, European integration, growing environmental awareness and a dynamic process of globalisation has been researched in detail and documented from a historical perspective for the first time.
- Against the background of the German Federal Government's digitisation strategy, the association has developed range of digital educational materials. A variety of interactive infographics designed specifically for use on whiteboards, PCs and tablets in schools has been created to explain the subject of adhesives to school students using modern teaching methods.

One of the challenges facing the industry is ensuring that an adequate supply of skilled employees is available in future. The publication of the newly developed teaching materials entitled "Die Kunst des Klebens" (The Art of Bonding), which have been distributed to almost 17,000 technology teachers throughout Germany, represents an important contribution by the German adhesives industry to the education of school students that brings the subject of bonding and adhesives to the attention of young people. The e-paper "Berufsbilder" (Job profiles) highlights for school and university students the varied career opportunities available in the German adhesives industry. The objective is to encourage the adhesives experts of tomorrow to become familiar today with the industry and the variety of options it has to offer.

At the General Assemblies in 2016 and 2017, the following companies became members of the German Adhesives Association:
- ARLANXEO Deutschland GmbH, Cologne
- BLUFIXX GmbH
- Chemische Fabrik Budenheim KG, Budenheim
- cph Deutschland Chemie Produktions- und Handelsgesellschaft mbH
- Drei Bond GmbH
- Emerell GmbH, Buxtehude
- Gößl + Pfaff GmbH
- Rocholl GmbH, Aglasterhausen
- Saint-Gobain Weber GmbH, Merdingen
- UNITECH Deutschland GmbH

Sustainability

Sustainability means preserving a system for the good of future generations. In other words, sustainable enterprise management does not focus on the short-term success of a product or service, but rather aims to make a positive contribution over the long term. At the same time, attention must be paid to ecological, social and economic aspects.

There have been many long discussions in IVK as to whether and how sustainability aspects can be used to create added value for society and the economy without taking action just for the sake of it and incurring unnecessary costs. However, now that politicians and standardisation committees have taken up the topic, the future direction has been set. Even individual sectors of the industry, i.e. large-scale building projects, occasionally demand to see proof of compliance with 'sustainable' building practices. The objective of resource-efficient building is specified in the building products regulation EU/305/2011 (BRCW 7).

The objective of sustainable building entails an integral analysis and optimisation of buildings to improve the quality of the environment, society and the economy throughout the buildings' entire life cycle. Currently, attention is clearly directed at assessing and evaluating ecological aspects of energy and resource consumption. A wide variety of organisations have already taken the field to determine how sustainability can be verified and certified. In Germany, DGNB is responsible for the certification of buildings, while BNB assumes responsibility for government buildings. In other countries, BREEAM, LEED, HQE and other systems have been established. Since the sustainability impact assessment of buildings is primarily based on the products used and their upstream products as well as related production processes, they must also be evaluated. For the analysis of life cycle assessments, the consulting company Thinkstep has the requisite information (e. g. about primary energy consumption and identified impacts, such as the greenhouse effect, ozone depletion, potential acidification and eutrophication) to draw up ecological assessments and perform life cycle assessments, which are an integral part of the EPDs (Environmental Product Declarations). In Germany, the Institut Bauen und Umwelt e. V. (IBU: Institute for Construction and Environment) plays a major role in the development and issuing of EPDs. IBU maintains an EPD programme that encompasses the preparation and review of details for completeness and plausibility and verification by an independent third party. IVK is a member of IBU and has been working closely with its sister associations Deutsche Bauchemie (German Industry Association for Construction Chemicals) and Verband der deutschen Lackindustrie (VdL: Association of the German Paint Industry) since 2010 to develop an industry solution for drawing up sample EPDs, which should help manufacturers to provide the necessary information for building certification. There are sample EPDs for a wide variety of building products, which also include construction adhesives and cover almost the entire range of segments. Sample EPDs and guidelines are now available for

- reaction resin products based on EP
- reaction resin products based on PU and SMP
- plastic-modified mortar and
- water-based products

These were verified by IBU and are published on the IVK website: http://epd.klebstoffe.com/. English-language versions are also available.

On the basis of this model, work is currently being carried out by FEICA to establish the sample EPDs on a European level: http://www.feica.eu/our-priorities/key-projects/epds.aspx.

Prominent certification bodies are supporting the project, which is also being promoted by ECOPlatform (http://www.eco-platform.org/) using its logo. The members of ECOPlatform are the leading European program operators. This will respond to the 'key requirement' of the building products regulation on sustainable building.

The US-American adhesives association ASC is also interested in this approach and has had detailed discussions with IVK on developing a PCR (product category rule), which is the precursor to an EPD for sealants. However, for cost reasons the ASC has stopped its work on sample EPDs for the time being.

For industrial sectors, the values for the **"Product Carbon Footprint"** (PCF) are more suitable. This is an attempt to reduce the sustainability assessment to a single value – the CO_2 balance of a product. This approach, which was in part initiated by politicians, has been rejected by many experts in the meantime. Indicating a PCR for adhesives as unfinished products makes little sense, since they are never used alone but rather always as a system component with at most a marginal impact on the PCF of the final product. A project that was conducted jointly with the Bundesverband Druck & Medien (German Print and Media Association) to determine the impact of adhesives on the CO_2 balance during the production of books has clearly shown that, depending on the type of adhesive used (dispersion, hot melt), the percentage that can be attributed to the adhesive is usually between 0.06 and 0.5% and is thus negligible when compared to paper usage. It can generally be assumed consequently that the CO_2 footprint of adhesives is very small compared to the products that are manufactured using them. Although the quantity of adhesive used in the manufacturing of products is usually very small, adhesive manufacturers are also asked about the PCF values of the adhesives that they supply. Many studies in recent years have shown that the PCF value of adhesives in their delivered state

(cradle to gate) can be summarised with a view to the error limits of PCF calculations within certain product groups. On the initiative of the Technical Committee, PCF values (cradle to gate) for the following product groups were therefore identified:
• Adhesives based on aqueous plastic dispersions
• Thermoplastic hot melt adhesives
• Solvent-based adhesives
• Reactive adhesives

The corresponding PCF documents and more detailed information on the individual product groups and the database can be found in the IVK information document "Typical Product Carbon Footprint (PCF) values for industrial adhesives". (http://www.klebstoffe.com/fileadmin/redaktion/ivk/M-RS_2014-19_Anl_Product_Carbon_Footprint___www__.pdf).

Showing that adhesive bonding represents "more sustainable" production solutions compared to alternative joining methods is much more important than indicating the PCF for an adhesive bonding system and confirming its marginal impact on the CO_2 balance of the final product. Eco-friendly lightweight construction in automotive engineering applications is an outstanding example of the positive influence that adhesive bonding has. Using epoxy resins to produce rotor blades for wind power plants is another such example, and clearly demonstrates how to improve the CO_2 balance by leveraging modern material technology. National efforts should be standardised and recognised on a European level. This offers the possibility of new market opportunities through the identification of sustainable solutions for selected applications, thereby eliminating the need to produce individual analyses. Examples can be found at http://www.feica.eu/our-priorities/sustainable-development.aspx.

Guidelines "Adhesive Bonding – the Right Way"

Fabricating durable and reliable bonds means more than just selecting the right adhesive. Important parameters that must be taken into consideration include, for instance, material characteristics, surface treatment, a design suited to bonding applications or proof of product reliability. In this context, the German Adhesives Association, together with the Fraunhofer Institute IFAM, has developed the interactive guidelines "Adhesive Bonding – The Right Way". The guidelines have been prepared specifically for skilled crafts and industrial enterprises that require additional information about adhesive bonding systems.

The use of adhesives is so multifaceted nowadays that it is not possible for adhesive manufacturers to take into account all applications and in particular specific areas of use in material data sheets. With the guidelines "Adhesive Bonding – The Right Way", IVK and IFAM have created a practical aid for skilled crafts and industrial enterprises which require both general and additional information. The planning, development and production of a fictitious product are explained step-by-step, making sure that all steps of the planning and production phase are systematically taken into account. The ability to perform all the necessary process steps and observing the correct sequence ensures a high degree of quality assurance that should not be underestimated, and the interactive guide is the right instrument in this regard. The guide also contains a glossary and a search function so that key points of practical adhesive bonding applications are thoroughly covered.

The guide can be used free-of-charge and interactively at the internet portal of the German Adhesives Association: http://leitfaden.klebstoffe.com

The guide is also available in English ("Adhesive Bonding – the Right Way") at: http://onlineguide. klebstoffe.com

Executive Board

The membership of the Executive Board of the German Adhesives Association reflects the existing corporate structure of the German adhesives industry, which consists of small and medium-sized businesses as well as companies operating at a multinational level. Furthermore, the balance in the board's members guarantees access to the highest level of core competence and expertise relating to the key market segments that are important to the adhesives industry.

A top priority for the Executive Board of the German Adhesives Association is to adapt the association's structure and its committees to new economic and technological constraints and conditions in a continuous and timely manner in order to ensure that the organisation always operates efficiently and provides the maximum benefit for the German adhesives industry.

Holding technical discussions, assessing the economic, political and technological trends in the various key market segments of the adhesives industry and monitoring and analysing the activities of the association's numerous commercial and technical committees form an integral part of the responsibilities of the association's Executive Board.

The German Adhesives Association is considered to be the foremost competence centre in the field of adhesive bonding and sealing. It is the world's largest national association and the leading body in terms of its comprehensive service portfolio for adhesive bonding technology. IVK's links with the German Chemical Industry Association (VCI) and its various technical sections form the basis for its successful position. In addition to its connections with the chemicals industry, the association has a strategic and highly-efficient 360-degree network of expertise consisting of all the relevant system partners, scientific institutions, leading trade and industry associations, employers' liability insurance associations, consumer organisations and the organisers of exhibitions, training courses and conventions. As a result, the association covers every single link in the adhesives value chain.

The growing importance of the adhesives industry within the network of the chemical industry is also reflected in the fact that the German Adhesives Association has had a total of four seats on the central committee of the German Chemical Industry Association since the spring of 2016.

Within the framework of the strategy developed by the association's Executive Board for a "qualified market expansion", IVK actively supports a wide variety of scientific research projects relating to the field of adhesive bonding technology. This systematic approach to research is primarily interdisciplinary.

In practical terms, this means that scientific knowledge is combined with engineering expertise in order to produce practical research results. Consequently, this approach has contributed to the fact that adhesive bonding has now become predictable and is firmly established as a reliable bonding and sealing technology in the field of engineering. It is now rightly and undisputedly considered to be the key technology of the 21st century.

As a founding member of the adhesive technology group under the umbrella of DECHEMA and the joint committee on adhesive bonding (GAK), IVK maintains regular contact with all the relevant research institutions involved in steel, timber and automotive research. Together, they jointly assess and support publicly funded scientific research projects in the field of adhesive bonding technology. This cooperation has enabled the German Adhesives Association to successfully establish important research projects with results that promise significant benefits in particular for the adhesives companies which are members of the association.

The employee training programme of the Fraunhofer IFAM in Bremen, which has received strong support from IVK in terms of finance and content, has evolved into a well-established, fully recognised educational programme that will continue to be developed further. More and more users of adhesives have realised the advantages of having their employees receive qualified training in the appropriate use of technically demanding bonding and adhesive systems. In turn, the adhesives industry has also benefited in every respect from working with competent professional system partners. After Germany's Federal Railways Office made it mandatory to employ qualified staff for the application of adhesives during the manufacture of rail vehicles, the automotive industry and other adhesives users decided to launch an initiative that also focused on improving the quality of their products and increasing efficiency in production by exclusively using employees who have been trained in adhesive bonding technology. The result of this initiative, which was supported by experts from the German Adhesives Association, is the DIN 2304 standard "Adhesive bonding technology – Quality requirements for adhesive bonding processes". This mandatory standard was published in the spring of 2016. It specifies the requirements for the creation of high-quality, professional structural and load-bearing adhesive bonds in specific safety-related areas. In addition, the standard requires knowledge of and training in the use of adhesives. The training of staff in adhesives use, which was a voluntary measure in the past, has now become a mandatory component of bonding processes. Businesses which manufacture products on the basis of this standard must document the fact that they are using the latest adhesives technology. Seen from this perspective, this initiative, which currently applies only in Germany, clearly also has a European or even a global dimension.

The Executive Board's strategy to establish an employee training programme within the European and international markets has been successful. The course contents, which were developed by IFAM with the financial support and expertise of IVK for the various levels of qualifications (adhesive bonder, adhesive specialist, adhesive engineer), have been translated into English, Chinese and other languages and adapted to the different applicable European and international standards. Training courses are now held on a regular basis for adhesive bonders in Poland, the Czech Republic, Turkey, the United States, China and South Africa. Over the past 23 years, more than 7,000 people have completed the employee qualification programme at the Adhesives Technical Centre of the Fraunhofer IFAM in Bremen, at TC-Kleben in Übach-Palenberg and in other locations in Germany and abroad.

By developing this global training standard and putting in place a suitable worldwide training programme, the Executive Board of the German Adhesives Association has clearly demonstrated the competence and expertise of Germany's adhesives industry on both the European and international markets.

For IVK's Executive Board, having sound training and qualifications in adhesive bonding technology is just as important as ensuring that young people have a good education in subjects such as chemistry, engineering and material sciences. Against this background, the contents and educational approach of the programme "Die Kunst des Klebens" (The Art of Bonding) have been comprehensively revised by experts from IVK, the GCIF and educational specialists with the financial support of the German Chemical Industry Association (VCI). The course material is available to almost 20,000 technology teachers all over Germany.

In addition to this teaching material, the German Adhesives Association, in cooperation with FWU, the Institute for Film and Picture in Science and Education, has designed and produced two educational DVDs. The DVD entitled "Grundlagen des Klebens" (Adhesive Fundamentals) was developed for classroom use in schools and vocational colleges, while "Kleben in Industrie und Handwerk" (Adhesives in Industry and Trades) describes specific practical applications of adhesives involving different combinations of materials and is ideal for use in engineering and materials science teaching at vocational colleges. Both DVDs consist of a variety of films, animated sequences, interactive assessment tests and extensive information for instructors and students. The DVDs are available online from the FWU media library at www.fwu.de.

Against the background of the German Federal Government's digitisation strategy, the association has worked together with the German Chemical Industry Fund to develop digital educational materials. A range of interactive infographics designed specifically for use on whiteboards, PCs and tablets in schools has been created to explain the subject of adhesives to school students using modern teaching methods. In addition, the association has commissioned the Institute of the Didactics of Chemistry at the University of Frankfurt to develop a teacher training programme on the subject of "Adhesives and bonding in teaching" in order to give a more in-depth insight into the chemistry of adhesives and to highlight links with the school curriculum.

The association is also supporting Germany's unique, nationwide development program for young people known as "Fraunhofer MINT-EC Talents", which aims to provide support for particularly gifted school students and to encourage them to study the STEM subjects (science, technology, engineering and mathematics). Scientists from the Fraunhofer IFAM run workshops on the subject of bonding for talented young people over a period of two years until they leave secondary school. The high levels of participation in the "Jugend forscht" research competition for young people and the excellent results it has produced clearly demonstrate the benefits of providing support at an early stage. The ground-breaking invention of a biodegradable plastic film made from natural, renewable raw materials was awarded the special chemistry prize in the German national final of "Jugend forscht 2016".

In addition to human resources development, qualifications and general education, IVK also helps its member companies with recruiting trainees and skilled professionals. In order to counteract the imminent shortage of skilled professionals that is the result of demographic changes, IVK's Executive Board has launched its training initiative for the adhesives industry. With the "komm kleben…" (Join the Bonding Community) campaign, Germany's adhesives industry aims to present itself to young people on social media as an industry with interesting career opportunities. It has developed and published its own e-paper entitled "Berufsbilder in der deutschen Klebstoffindustrie" (Job profiles in the German adhesives industry).

IVK is also increasingly becoming a leading player both in Europe and worldwide with regard to technical issues.

This applies in particular to the concept, jointly developed by the Executive Board and the Technical Board, of a standardisation competence platform in the German Adhesives Association. The objective behind this project is to make use of the in-depth involvement of the German Adhesives Association in European standardisation (CEN) and to play an active role in international standardisation activities at the ISO level. The main factors behind this initiative were, on the one hand, the increasing number of ISO standards for adhesives which are gaining growing acceptance as part of the globalisation of markets. On the other hand, the Executive Board is pursuing the goal of introducing important German industrial standards to the global adhesive industry in the long term. With its standardisation competence platform, the association has succeeded in setting up and implementing three important standards projects in the fields of adhesives for floor coverings and wood on an international level. In addition, it has provided key information to member companies about future developments in the electronics industry and the requirements for the relevant adhesives.

The German Adhesives Association has taken responsibility for managing the European standardisation project "Mandated flooring adhesive standard" and the European Standards Secretariat "Wood and Wood Materials". These projects will ensure that the interests of Germany's manufacturers of construction adhesives have adequate representation with respect to European requirements on indoor air pollution, with the aim of avoiding the need to comply with the German Institute for Building Technology's (DIBt) indoor air pollution requirements, which involve a considerable amount of bureaucracy. The head of the Standards Secretariat for Wood and Wood Materials will protect the specific interests of the German adhesives industry in the complex regulated market for glued laminated timber for load-bearing applications.

In order to be able to take a practical approach to sustainability, a subject currently being given a high priority by politicians and the business world, on behalf of the German adhesives industry, the German Adhesives Association is continuing to focus on a variety of projects:

In cooperation with several sister associations in the German Chemical Industry Association (VCI), an industry solution has been drafted for generic environmental product declarations (EPDs).

The principles of sustainable buildings are governed by the EU Construction Products Regulation. EPDs can be used to prove that construction products contribute to the creation of a sustainable building. They include detailed analyses and documentation (CO_2 consumption in production, energy consumption, environmental assessments, life cycle analyses etc.). In consultation with the German Association for Sustainable Building (DGNB), over 150 different adhesive-related sample EPDs in four product categories have been produced for construction-products, covering a number of product families, a variety of raw materials and most construction applications. The industry-specific EPDs create a standard framework for the industry concerned and can be used by all the members of the association affected. The German Adhesives Association has made the EPDs and the accompanying guidelines available to the industry in German and English on its website www.epd.klebstoffe.com. At the insistence of IVK, FEICA has developed and published sample EPDs for European use on the basis of the German sample EPDs. They are available at http://www.feica.eu/our-priorities/key-projects/epds.aspx. These sample EPDs are gaining increasing acceptance in Europe and this confirms the success of the approach taken by IVK, despite the fact that some national sustainability certification systems require additional information. The EU Commission is working on developing EN 15804, the basic standard for EPDs, for the purposes of harmonisation.

The US Adhesive and Sealant Council (ASC) also indicated its interest in introducing sample EPDs to the USA in order to provide a useful service to its members. IVK provided expert support to ASC with the aim of bringing about a global standard for environmental product declarations. However, after a product category rule (PCR) had been drawn up for sealants, the work came to a standstill following a survey, because the costs would have to be borne by the members with an interest in this area. It is not clear whether the work will restart at a later date.

A parallel project that evolved from the EPD project focused on calculating PCFs (Product Carbon Footprints) for individual families of industrial adhesives. This is important information for the adhesives industry, which will allow sustainability statements for adhesives to be produced.

An project that was organised jointly with the Bundesverband Druck & Medien (German Print and Media Association) to assess the impact of adhesives on CO_2 consumption during the production of books clearly shows that the percentage that can be attributed to the adhesive is between 0.06 and 0.5 %, depending on the type of adhesive used, and is therefore negligible. As a result, it can generally be assumed that the CO_2 footprint of adhesives has only a marginal impact on CO_2 consumption during the production of glued products. Additional practical examples will be collected on an ongoing basis to substantiate this assumption.

Regardless of the results of these additional investigations, it is clear that innovative, modern adhesive systems are essential for producing eco-efficient products. This applies both to light-weight construction in the automotive and furniture industries and to the manufacturing of products used to generate renewable energies, such as solar cells and wind turbines.

The Executive Board believes that the circular economy is a new and very important topic that will have a significant influence on the demands which will be imposed on adhesives and adhesive processes in the decades to come. The European Ecodesign Directive already specifies the requirements for bonded joints in electronic displays, servers and computers. The German adhesives industry supports the objectives of the circular economy package and underlines the specific importance of the use of innovative bonding solutions to achieve the goals of reuse, recycling and repair. The Technical Board is coordinating support for the legislative process for the circular economy package on a national and European level and is ensuring that requirements relating to adhesives are given sufficient consideration.

The fact that the German Adhesives Association and its members actively participate in important global conferences underlines the outstanding international position of the German adhesives industry. This applies in particular to the International Adhesive & Sealant Conference that takes place every four years. Following the international conference in Paris in 2012, the German Adhesives Association supported the 2016 International Adhesive & Sealant Conference in Tokyo with a number of interesting presentations. Germany's adhesives industry and its association are therefore actively taking the opportunity to highlight the technological leadership of the German adhesives industry to an international audience.

During the Executive Board's regular interactions with adhesives associations in the United States and Asia, it is becoming increasingly evident that the German adhesives industry is not only considered to be a global technological leader but is also regarded with respect for its competence in "Responsible Care®" and sustainable development. In consultation with its system partners, several years ago the German adhesives industry began to develop practical solutions to provide adequate protection for the environment and consumers and ensure workplace safety, before the relevant legislation was introduced, and went on to successfully implement these solutions in line with the strategic guidelines of the board. By separating solvent consumption from the production of adhesives and successfully establishing the EMICODE® and GISCODE systems and sample EPDs, the German adhesives industry has fulfilled its responsibility to the environment and its customers throughout the entire value chain and is therefore playing a leading role on a global scale. With voluntary initiatives to remove solvents from parquet adhesives and phthalates from paper/packaging adhesives and with its information series "Klebstoffe zur Herstellung von Materialien und Gegenständen, die dazu bestimmt sind, mit Lebensmitteln in Berührung zu kommen" (Adhesives used in the manufacturing of materials and articles intended to come into contact with food), the German Adhesives Association has repeatedly set new standards in terms of health and safety at work and environmental and consumer protection.

The Executive Board of IVK believes that maintaining a balance in the relationship between technological leadership and social competence is the key to successful and credible positioning in the industry. This position ensures that the association always has access to reliable and important information, for example relating to the future direction of draft legislation, and that members of the association will have the opportunity in the future to open up new markets in Europe and worldwide with products that comply with the relevant requirements on workplace safety and environmental and consumer protection.

The members of the Executive Board view the constantly growing number of participants at the various events organised by the German Adhesives Association and the nine new member companies which have joined the association during the past two years as a clear indication that the German Adhesives Association is well positioned, provides a valuable and very practical service for the benefit of its members and is therefore highly attractive to the adhesives industry as a whole.

Technical Board (TA)

In its regular meetings in 2015 – 2017, the Technical Board focused on the specialist work of the technical committees, the subcommittees and the ad-hoc committees, discussed this in detail and developed scenarios for the adhesives industry. Furthermore, in-depth discussions were held on a number of interdisciplinary topics and issues which were then proactively influenced by the association's corresponding activities.

A major priority of the Technical Board's work has been sustainability. For instance, the Technical Board focused on issues such as how to define sustainability with regard to adhesive technologies and also ran specific projects on this subject.

In consultation with FEICA, the Technical Board emphasised the special role of adhesives as an enabler for sustainable solutions. However, the Technical Board considers the description of an adhesive as a "green adhesive" to be nonsense, since it does not provide any technical information. The Technical Board moreover thinks that environmental marketing of an adhesive based on individual indicators, for example its PCF (product carbon footprint), has no scientific basis. The environmental footprint of an adhesive should always be considered within the context of the respective application. Although the quantity of adhesive used to manufacture products is usually very small, adhesives manufacturers are nevertheless often asked for the PCF values of the products they supply. Numerous analyses carried out in the last few years have shown that the PCF value of adhesives as delivered (cradle to gate) can be combined within certain product groups with regard to the tolerances of PCF calculations. Therefore, on the initiative of the Technical Board, PCF values or bands (cradle to gate) were created for four different product groups (water-based dispersion adhesives, thermoplastic hot melt adhesives, solvent-based adhesives, reactive adhesives) and made available to customers in the form of an information sheet.

Another important project is the development of environmental product declarations (EPDs) for the building industry. This work was necessary as the EU Construction Products Regulation has called for the use of "sustainable" products since 2011 and the sustainability of products must be substantiated by EPDs. The legally required EPDs encompass detailed analyses and documentation, for example with regard to CO_2 consumption in production, energy consumption, resource depletion, ecological assessments, life cycle assessments etc. The preparation of EPDs involves considerable work and substantial costs, which means that an industry-wide approach is the ideal solution. The national (German) EPDs have been completed and have been offered to other European countries by FEICA. The development of FEICA EPDs on the basis of the German EPDs has now been finished. The ecological assessment data for the products covered by the EPDs has only changed slightly. In addition, FEICA has been a member of ECOPlatform since the start of 2016. The objective of ECOPlatform is to standardise and network all the existing EPD programmes in Europe. The meetings with ECOPlatform and the European Commission have been very positive and an agreement was reached on the mutual acceptance of EPDs and ecological assessments. The US Adhesive and Sealant Council (ASC) also indicated its interest in introducing sample EPDs to the USA. IVK provided expert support for ASC in order to bring about a global standard for environmental product declarations. However, after a product category rule (PCR) had been drawn up for sealants, the work came to a standstill following a survey, because the costs would have to be borne by the members with an interest in this area. It is not clear whether the work will restart at a later date. The international harmonisation of EPDs remains a challenge.

Another focal point of the Technical Board's work over the past two years included the activities relating to Europe's chemicals legislation "REACH" (Registration, Evaluation, Authorisation and Restriction of Chemicals), set out in the REACH Regulation (EU) No 1907/2006.

A central element of this regulation is the requirement for the "safe" use of chemicals across the entire supply chain. As a result, the duties and responsibilities have been shifted more towards downstream users. Proof of the safe use of their raw materials causes great problems in particular for the manufacturers of raw materials when they register the raw materials that they use to produce adhesives. On the other hand, the adhesives manufacturers have great difficulties drawing up use scenarios and safety data sheets for their adhesives that comply with REACH as a result of the unclear and unsystematic REACH requirements for mixtures. The reason behind these problems is that many adhesives contain 20 - 30 ingredients and, for many of these raw materials, there are extended safety data sheets, frequently with hundreds of pages. The unstructured contents of these data sheets have to be examined before they are included in the safety data sheet for the adhesive on a product- and application-specific basis. For this reason, the Technical Board is supporting a FEICA working group which is developing special solutions in this area. The ECETOC-TRA model used by the manufacturers of raw materials and others calculates user exposures and, in particular, consumer exposures extremely conservatively. As a result, DIY and other applications of adhesives are often formally "unsafe". This situation can be improved by means of "Sector-specific Consumer Exposure Determinants" (SCEDs) and "Sector-specific Worker Exposure Descriptions" (SWEDs) which represent much more accurately the actual exposures during the use of adhesives. The standard "Environment" model also produces highly conservative calculations using the prede-fined Environmental Release Categories (ERCs). Here too, many adhesives applications are formally "unsafe". This situation can be improved by means of "Specific Environmental Release Categories" (SpERCs), which, like the SCEDs and SWEDs, have now been published and are available to all registrants free of charge.

In the "Generic Exposure Scenarios" (GES) project, generic exposure scenarios are developed for groups of adhesives (solvent-based, reactive and cementitious adhesives). Generic exposure scenarios allow downstream users to draw up a "Downstream User Exposure Scenario" of their own for a chemical safety report according to REACH Art. 37(4), for example if the supplier exposure scenario is not effective and is not adapted by the supplier.

IVK members receive information relating to REACH via the VCI's REACH portal as well as in the newsletters published by IVK to keep its members informed about adhesive-related issues and the seminars it conducts on a cooperative basis.

To further raise the general public's awareness of adhesives as a joining technology, the guide "Sichere Herstellung von Klebungen" (Safe production of adhesive bonds) was drawn up together with IFAM. The guide "Kleben – aber richtig" (Adhesive Bonding – The Right Way) is now available on the internet both in German and in English.

In the past two years, IVK has further strengthened its commitment to standardisation activities in DIN, CEN and ISO. In January 2013, the new standards committee "Adhesives; Test Procedures and Requirements" was founded at DIN and in future this will function as an umbrella committee for the DIN standards committees that deal with adhesives. The founding members were representatives of the committees dealing with adhesives at the time. The new committee NA 062-04-59 AA has taken on the coordination of the "Adhesives" committees and has acted as a mirror committee of CEN/TC 193 "Adhesives". As a result, new subcommittees can be set up relatively quickly in future if required, making it possible to react to new requirements more rapidly than before. Dr. Udo Windhövel is the Convenor of NA 062-04-59 AA and Ansgar van Halteren is his deputy.

Efforts were made at ISO level to speed up standardisation processes as much as possible. For example, the option of bypassing individual standardisation stages has been introduced. This involves both risks and opportunities. On the one hand, the chance of raising an objection could be missed. On the other hand, it is possible to significantly accelerate the process by bypassing standardisation stages.

The large number of national, European and international standards relating to adhesives makes it almost impossible for adhesives manufacturers and, in particular, adhesives users to identify the relevant standards for a specific adhesives-related issue. Therefore, the Technical Board has launched a project with the aim of listing and categorising all the standards relating to adhesives to offer adhesives users a tool that they can use to find the standards relevant to a specific issue among the many standards already in existence. The contents of almost 800 documents have been described and the documents have been categorised on the basis of various different criteria, including the adhesive type, the application and the results of the test method. The list contains DIN, EN, ISO and ASTM standards, DVS guidelines and documents produced by other organisations (VDA, VDI, SAE, IVK etc.). The tool consists of a database with different filter groups (adhesive status, adhesive type, substrate, objective/results of the test and others) which allows the standards relevant to a particular issue to be filtered online. The database can be accessed via the IVK website and is regularly updated.

Because isocyanates can cause sensitisation of the respiratory tract, the Federal Institute for Occupational Safety and Health (BAuA) has carried out a risk management option analysis (RMOA) of diisocyanates as part of the REACH evaluation. The goal is reduce the number of occupational asthma cases caused by the use of isocyanates. Although the criteria for authorisation (Annex XIV, REACH) may have been met without a CMR classification ("chemicals of equivalent concern"), the question is whether the goal could be achieved much more efficiently with a restriction (Annex XII, REACH). While the BAuA favours this approach, activities are taking place in Sweden with the aim of achieving an authorisation. It is therefore in the industry's interest to actively support the BAuA in its efforts to put in place a restriction. The industry activities coordinated by the isocyanate manufacturers' associations ISOPA and ALIPA directly involve FEICA and IVK, which has set up its own ad-hoc group for this purpose. The BAuA officially submitted the restriction dossier on 6 October 2016. After the Committee for Socio-Economic Analysis (SEAC) and the Committee for Risk Assessment (RAC) had given their initial opinions, the six-month public consultation period began on 22 March 2017. The restriction is expected to come into force in 2019 at the earliest. The planned restriction will result in a ban on the sale and the professional and industrial use of isocyanates and mixtures containing isocyanates, with the exception of mixtures with a cumulative diisocyanate content that is less than 0.1% and of safe product usage combinations that result in only an insignificant risk or where users can prove that they have undergone the required training. As part of the cooperation with the BAuA, the ad-hoc IVK group has taken part in technical meetings with the consultancy company commissioned by the BAuA to evaluate the socio-economic effects of a possible restriction on the use of diisocyanates. The group is currently working as part of an industrial consortium to identify safe product usage combinations and prepare the content of future training courses.

After the German Environment Agency (UBA) withdrew its guidelines and positive lists in 2005, manufacturers of anaerobic thread sealants no longer had any means of obtaining the drinking water approvals (KTW certificates) required by the market. The approach taken by testing institutes at the time, which involved using testing methods that bore little resemblance to practical applications, coming to approximate conclusions about the influence of the cured adhesives on drinking water and then producing test certificates based on the guidelines that had been withdrawn, was highly unsatisfactory. Therefore, in 2013 IVK set up a working group consisting of the affected adhesives manufacturers. This group collaborated with the agency to establish a valid legal framework and develop a practical test which would allow for fair competition and ensure customer satisfaction.

As the UBA has estimated that it will take four to five years to produce a revised drinking water authorisation for anaerobic adhesives, the working group requested a written statement from the agency which will make it clear that, because of a lack of evaluation criteria, KTW approvals cannot currently be issued for these products.

The agency has now introduced the group of "marginal products" which includes anaerobic adhesives, primarily as a result of the close cooperation between the ad-hoc IVK group and the agency. An important factor is that the raw materials used in these products do not need to be evaluated in this case, which means that there is no longer a positive list for marginal products. In addition, there are no formulation-related requirements for individual materials or other additional requirements. However, the thread sealants must follow the outline formulation guidelines agreed with the agency.

The German Environment Agency has published the information document "Hygienic assessment of anaerobic adhesives that come in contact with drinking water" on its website: (http://www.

umweltbundesamt.de/themen/wasser/trinkwasser/trinkwasser-verteilen), but the document is only available in German. The agency does not consider a certificate of conformity to be necessary for anaerobic adhesives. The consistent quality of the formulation of anaerobic adhesives and the traceability of these products can be guaranteed by the manufacturers' own quality management systems. There is also the option of having an independent body verify that the formulation of an anaerobic adhesive corresponds with the formulation guidelines. IVK will make a separate logo available to its members for this purpose. This means that anaerobic thread sealants are once again covered by the provisions of legislation on drinking water, but with much less red tape and much lower costs.

There has been and still is a need for action with regard to the European Biocidal Products Regulation (EU Regulation No 528/2012). As part of its evaluation of active biocidal substances that are already in use, the EU Commission is issuing approvals for in-can preservatives in the form of implementing regulations. In particular in the case of substances with sensitising properties, the annex of the implementing regulations for product type 6 (in-can preservatives) generally contains a clause that makes mandatory the additional requirements for labelling specified in Art. 58(3) of the Biocidal Products Regulation 528/2012. The decisive factor in determining the deadline for implementing the labelling requirements is the date of approval of the substance. Only then do the special provisions of the implementing regulation come into force. IVK has informed its members in a number of newsletters about the approval of various in-can preservatives, the resulting changes in labelling and the accompanying deadlines. As an additional service, it has made available a table on the IVK intranet which gives an up-to-date overview of the in-can preservatives that have been newly approved by the EU Commission. The table indicates whether the implementing regulation for the active substance used as an in-can preservative in adhesives contains new labelling requirements in accordance with Art. 58(3) of the Biocidal Products Regulation 528/2012 and when these must be implemented.

On the basis of an initiative by the German Federal Ministry for the Environment (BMUB) to reduce VOCs due to the risk of summer smog, VCI was asked, together with other industrial associations, to explore the possibilities for reducing VOCs. Driven by concern for the environment, this initiative has led adhesive manufacturers to reduce their solvent consumption significantly over the past few years, as can be seen in the IVK solvent statistics assessment. The Technical Board has addressed this topic in depth and was able to demonstrate with the aid of its solvent consumption survey that the objective of reducing VOCs by 70 % by 2007 (based on the consumption figures in 1988) was in fact reached much earlier. The solvent statistics are compiled every two years and used in discussions with German and European agencies and institutions, for example when introducing new legislative proposals. The statistics are a very important instrument for communicating with agencies and authorities and are indispensable for documenting the environmental awareness of Germany's adhesives industry.

On 23 March 2017, Regulation (EU) 2017/542 on harmonised information relating to emergency health response appeared in the official journal of the European Union as Annex VIII of the CLP regulation. The regulation covers the harmonised reporting to the bodies appointed by the member states (in Germany: the Federal Institute for Risk Assessment) of information on the composition and other product features of mixtures that have been classified as harmful to human health according to the CLP regulation. Information must also be provided about mixtures that are only categorised as harmful because of their physical and chemical properties. Mixtures that are classified as being harmful only to the environment do not need to be reported. The products that are reported are indicated by a UFI (unique formula identifier) on the label. In the case of products not intended for consumers, the UFI can be included only on the safety data sheet. Every change in the formulation which results in the product being reported again will require a new UFI. In the report, all the known components of the formulation (>0.1 % with critical hazard classifications and >1 % without critical classifications) must be specified in narrow classification bands with the exact chemical name and CAS number.

The new EU regulation will be implemented gradually. The reporting of mixtures intended for end consumers in accordance with the harmonised requirements will be mandatory from 1 January 2020. For mixtures in professional use the date is 1 January 2021 and for mixtures only in industrial use it is 1 January 2024. The information about the contents of the mixtures from the safety data sheet is sufficient for mixtures only in industrial use, if there is a 24/7 information desk available.

Many questions about the practical implementation of the regulation remain unanswered and are being discussed in newly established working groups in ECHA. The question of the interpretation of "industrial use", in other words, in which cases a mixture is considered to be only in industrial use, has not been finally resolved and is currently under discussion in the relevant EU expert group (CARACAL). The regulation will involve a large amount of additional work and costs for the industry. All formulations must be checked to determine the classification band of the raw materials used and, if necessary, modified. The products must be grouped into consumer, professional or industrial mixtures. All the data concerning the categorisation and labelling of the mixtures that require reporting must be updated and UFIs must be generated. In addition, a monitoring process needs to be developed and the company's own software adapted in order to ensure that all the information on changes in the product portfolio is incorporated and, if necessary, that new or updated reports are sent automatically to the poison centres. The German Chemical Industry Association (VCI) and its professional bodies have developed guidelines that give an overview of the new obligations under EU law, the status of regulations in Germany up to 2020, the questions that are still open and the opportunities for companies to make preparations.

The European Commission is increasingly focusing on resource and material efficiency in its plans for the implementation of the ecodesign directive. This applies in particular to servers, PCs and displays/TVs. Other product groups, such as white goods, will follow. As progress is made with implementing the ecodesign directive in areas that extend beyond the operating phase of energy-related products, the emphasis is increasingly being placed on considerations of reusability, recyclability and repairability. This is reflected in documents relating to the revision of the relevant regulations for specific product groups. The proposals include requirements for disassembly and the publication of instructions etc., if necessary accompanied by a ban on welded and bonded joints.

The Technical Board believes that the subject of the circular economy and ecodesign is highly important and will have a significant impact on bonding processes over the decades to come. In a series of meetings with the German Federal Ministry for Economic Affairs and Energy (BMWi), the Federal Ministry for the Environment (BMUB), the Federal Institute for Materials Research and Testing (BAM) and the Environment Agency (UBA), IVK has made it clear that "bonding" does not prevent a product from being repaired or recycled (end of life). ("Debonding" must be one of the requirements placed on product manufacturers and must be taken into consideration during the product design process, in collaboration with the adhesives supplier.) IVK also explained that only a formulation which is neutral in technological terms (the specification of objectives instead of joining methods) will allow innovations to be developed in future. German and English versions of an IVK position paper on this subject in response to the EU Commission proposal "Ecodesign requirements for electronic screens" have been submitted to the ministries.

IVK will closely follow the implementation of the ecodesign directive for electronic and electrical devices in cooperation with FEICA and other associations and will continue to monitor the legislative procedure for the circular economy package and for the outline circular economy act in relation to its relevance to adhesives.

At the end of February 2016, the DIN standard 2304 "Adhesive bonding technology – Quality requirements for adhesive bonding processes – Part 1: Adhesive bonding process chain", which IVK played a significant role in developing, was published with a date of March 2016. This represents another important means of safeguarding the quality of the bonding process chain. The new standard specifies the requirements for the high-quality professional production of constructional/structural/load bearing bonded joints in pre-defined safety classes. The definition of the safety classes is based on the possible risks of injury or death in the event of the mechanical failure of a bonded joint. The standard does not cover the requirements for bonded joints with regard to suitability for use with food, fire prevention or compliance with emissions regulations or health and safety provisions. In addition to the documentation of the bonding process, the standard also requires knowledge of and training in the use of adhesives. This means that the training in adhesives on the basis of the safety class of the bond, which in the past was voluntary, has now become a mandatory component of controlled bonding processes. The tasks and responsibilities of bonding supervisors are specified in DVS guideline 3311. Companies that manufacture products in future in accordance with this standard must also document the fact that they are using the latest bonding technology. As DIN 2304 is a universal basic standard, more detailed information is needed in order to be able to apply it in practice. The additional information relates to the general part of DIN 2304, but details of individual materials and/or industries are also needed (for example for fibre composites or specific industrial applications). This information will be provided in a DIN specification (DIN Spec 2305-x). These can be created, adopted and published quickly and without a large amount of formal work. DIN has organised a translation of the specification into English following a number of requests.

The Technical Board has also focused intensively on the topic of "adhesives training". It strongly recommends taking steps to support and promote the training of adhesives users for the purpose of a qualified market expansion at a European level. In this context, the Technical Board supports a project by Fraunhofer IFAM to establish a certified training programme for "Adhesives Engineers" on an international level.

Since the appearance in 2002 of the first series of slides entitled "Bonding", which was produced in collaboration with the German Chemical Industry Fund, the subject of adhesives has fortunately been incorporated into the science syllabus and, in particular, the chemistry syllabus for secondary schools.

As a result of significant demand, the comprehensive teaching materials have been completely revised and brought fully up-to-date in terms of both content and teaching methods, once again in collaboration with the German Chemical Industry Fund.

The new teaching materials on the subject of "Die Kunst des Klebens" (The Art of Bonding) were distributed to almost 17,000 technology teachers throughout Germany at the end of 2015. Outside the school classroom, these materials are an excellent means of supplementing the content of company training courses and presentations.

"Die Kunst des Klebens" can be ordered from the IVK website, the IVK central office or direct from the German Chemical Industry Fund website at http://www.vci.de/fonds/unterrichtsmaterialien.

Other topics discussed by the Technical Board included:
• Involvement in technical regulations for hazardous substances (TRGS)
• Classification and labelling issues
• Health and safety at work, environmental and consumer protection issues
• Monitoring various EU Commission projects in Germany

Technical Committee Building Adhesives (TKB)

Introduction
The Technical Committee Building Adhesives (TKB) of the German Adhesives Association (IVK) represents the interests of manufacturers of building adhesives and dry mortar systems who are members of IVK. The committee liaises with public authorities, trade bodies, employers' liability insurance associations, other industrial organisations and standardisation committees.

Its aims are to establish technical standards, to influence the provisions of legislation on chemicals, to participate in developing legislation, to promote technical progress while safeguarding users and the environment, to provide technical support and information to customers and the building trade, and to promote the use of building adhesives and mortar systems by providing objective technical information.

Overview of topics
TKB's activities can be broken down into several categories:
• Technical topics relating to building adhesives/flooring installation products and their applications.
• Standards for flooring and parquet adhesives, levelling compounds, tile adhesives, binders for floor screeds, primers and special products.

- Technical information events for the flooring and parquet laying trade and other related trades.
- TKB publications about current topics relating to application methods and building legislation and about standardisation and environmental and user protection issues.
- Building legislation issues, including approvals in Germany and Europe.
- Topics relating to chemicals legislation, such as German, European and international regulations on labelling hazardous materials.
- Health and safety at work and environmental and consumer protection.

The committee's work
Mortar systems
On the national and international standards bodies ISO/TC 189/WG3, CEN/TC 67/WG 3, CEN TC 303, WG 2, NA 062-10-01 AA and NA 005-09-75 AA, representatives of TKB help to define the standards for flooring/parquet adhesives, tile adhesives, levelling compounds, sealing compounds and floor screed binders.

The requirements standard (EN 12004-1) and the test standard (EN 12004-2) for tile adhesives were adopted and published in April 2017. This brings them into line with the corresponding ISO standards (ISO 13007-1 and ISO 13007-2).

The revision of EN 13888 (Grout for tiles – requirements) will include the requirements for fast-drying cementitious tile grouts, in order to ensure that it corresponds with ISO 13007-3. Because EN 13888 is not a harmonised standard, the chapter on conformity assessments will be deleted. In a similar way to the approach for tile adhesives, the test methods for tile grouts from the EN 12808 series will be summarised in a new EN 13888-2 standard.

The situation for sealing compounds remains complex. Depending on the area where they are used, the proof of usability can take three different forms: either compliance with EN 14891 or ETAG 022 or a general building authority test certificate (abP) for sealants. The joint initiative with the German Institute for Building Technology (DIBt) to bring together system tests for sealing compounds and the accompanying system components in EN 14891 was rejected on a European level.

As a result of the European Court of Justice (ECJ) judgement C-100/13, the DIBt has withdrawn the construction products lists and introduced measures for the future regulation of buildings, rather than of building products as in the past. In future it should be possible, via the indirect route of building regulation, to maintain the supposedly high level of protection and therefore to avoid the consequences of the ECJ judgement. In the view of TKB, this is not compatible with the requirements of the Construction Products Regulation, because ultimately it will lead to additional national requirements being introduced for products that have already been CE marked.

The revision of DIN 18157 parts 1 – 3 (Execution of tiling by the thin mortar bed technique) has been completed and the revised standard was published in April 2017. The standard gives a threshold of 0.3 CM% for the readiness of heated calcium sulphate screeds for the installation of floor coverings, which differs from the value of 0.5 CM% in DIN 18560-1.

The work on the sealant standards in the series DIN 18531 to DIN 18535 has been completed. Their publication is planned for the third quarter of 2017.

Product and application technology

TKB continued its comprehensive investigations into the subject of screed drying, the readiness of screeds for the installation of floor coverings and subfloor moisture measurement methods. The fundamental information on determining the readiness of screeds for the installation of floor coverings by measuring the relative air humidity above screed samples is documented in TKB report 1. The description of the method (KRL or corresponding relative humidity method), including scientific background information, has been published in TKB report 2. In order to promote the use of the KRL method in practice, moisture sensors that are suitable for real-life measurement conditions have been identified and recommended. The results have been made available to the industry in TKB report 3. TKB is promoting the practical testing of the KRL method in collaboration with well-known experts. A joint measurement campaign has been launched in which CM and KRL moisture levels are measured in parallel on building sites. The results of the campaign are expected to be available at the end of 2017. The medium-term objective is to transform the existing TKB guideline values into recognised guideline values for the readiness of cement-based screeds for the installation of floor coverings. Alongside this work, TKB has published details of the latest technology for CM measurement in TKB technical information sheet 16. This technical information sheet is supported by all the main trade associations and, among other things, it specifies the moisture thresholds that have become well-established across the industry. In addition, legal information about liability when assessing the readiness of screeds for the installation of floor coverings has been included to ensure that companies laying floor coverings use a legally valid procedure for assessing the subfloor.

Standards

In cooperation with the IVK standards competence centre, TKB has helped to define European standards as part of the ISO/TC 61/SC 11/WG 5, CEN/TC 193/WG 4 and NA 062-04-54 AA standards bodies. The work on extending mandate M/127 to convert EN 14293 (Parquet adhesives) and EN 14259 (Adhesives for floor coverings) into harmonised European standards is still underway. The scope of EN 14259 is to be extended to include dry adhesives and fixings. The aim of extending the standards in this way is to avoid the need for the DIBt to issue supplementary national regulations for these product groups in the medium term. In a similar way to ISO 17178, which is based on the classification of parquet adhesives drawn up by TKB, EN 14259 will also be converted into an ISO standard. A separate requirements standard will not be provided for levelling compounds. Instead, TKB representatives are involved via NA 005-09-75 AA and CEN/TC 303 in revising EN 13813, which plays an important role in the CE marking of levelling compounds. Because of the specified deadlines, the revised standard must be published by August 2018. Cooperation between CEN/TC 193, which drew up the test standards for levelling compounds, and CEN/TC 303 is guaranteed by the membership of the technical bodies (liaison). Despite objections by TKB, the CM method has been integrated into part 1 of DIN 18560. At the same time, the recommended moisture threshold for heated calcium sulphate screeds of 0.3 CM% has been increased to 0.5 CM%. In the opinion of TKB, this will lead to a significantly increased risk of moisture damage. This view is shared by all the trade associations for companies that install floor coverings and parquet. As a result, TKB has published details of the latest technology for CM measurement in TKB technical information sheet 16 as a counterbalance to the standard. In practice, companies installing floor coverings, parquet and tiles will continue to use the existing threshold of 0.3 CM% for determining the readiness of heated calcium sulphate screeds for the installation of floor coverings. EN 1372, EN 1373, EN 1902 and EN 1903, which are important test standards for adhesives for floor coverings, have been revised. The processing of floor covering and parquet adhesives has been adequately described in standards, technical information sheets and books. In order to avoid duplicating the regulations in DIN 2304 "Quality requirements for adhesive bonding processes", a passage has been inserted in this standard which excludes adhesive processes that have already been adequately described elsewhere.

Publications / events

The 32nd TKB conference in 2016 and the 33rd TKB conference in 2017 covered topics from the fields of product technology, building legislation, health and safety at work and the environment. The subjects of the presentations included the assessment of the KRL measurement method; the combined CM/KRL measurement approach; the possibilities of decorative levelling compounds; the advantages of bonded floor coverings; laying floor coverings on stairs without the use of solvents; validating the odour testing of construction products; regulating construction products; measuring the moisture levels of subfloors using a new type of sensor; the manufacturing, suitability and installation of solid floors; laying floor coverings on system floors; the quality requirements relating to the smoothness of subfloors; the special features of the hardening process of silane-based adhesives and the creation and legal significance of standards.

TKB technical information sheet 10 has been revised and information about precast screeds made from gypsum fibreboard has been added. In TKB technical information sheet 16 the CM method is described as the generally recognised code of practice that is supported by the vast majority of the trades that lay floor coverings. TKB technical information sheet 15 on the subject of laying patterned PVC floor coverings and multi-layer flooring has been revised and published, after consultation with the relevant associations. TKB technical information sheet 1 Installation of parquet has also been revised. This information sheet was also made available for the last time in paper form. In future the TKB technical information sheets will only be available to download from the IVK website. In order to make the creation of TKB technical information sheets more transparent, the consultation process with the relevant associations has been amended. In future all TKB technical information sheets will be revised on a five-year cycle to ensure that they remain up-to-date. A specially created literature database guarantees that the source material for all the TKB technical information sheets is up-to-date and that the material is quoted in the same way across all the sheets.

Statements from TKB on current subjects, including new and revised TKB technical information sheets, the revision of test standards for adhesives for floor coverings and the ECJ blocking the measures taken by the DIBt were published in the industry press and on the IVK website under the heading "TKB informiert ..." (Information from TKB ...).

Cooperation with other associations / institutions

TKB consults with a variety of associations and organisations on technical and regulatory issues. The traditional TKB round table talks known as "Handwerk and TKB" (the trades and TKB) have been replaced by the "TKB Branchengespräch" (TKB industry meetings), which involve representatives of floor covering manufacturers. The subjects covered included the possibilities for categorising the quality of levelled surfaces; the trend for the loose and floating installation of patterned PVC floor coverings and multi-layer modular flooring; the inclusion in screed standards of the CM measurement method and the modified threshold for determining the readiness of heated calcium sulphate screeds for the installation of floor coverings; the references to the screed standard DIN 18560 in DIN 18365 (flooring work); the involvement of association representatives on standards committees; the ongoing development of the KRL measurement method; developments in the odour testing of construction projects; the importance of plasticisers in parquet adhesives and the reclassification of floor laying materials containing isocyanates.

TKB has shared information with the German Federal Association of Screeds and Floor Coverings (BEB), in particular on the subject of screeds. Key issues included consultation on technical information sheets concerning the preparation of subfloors, the latest findings on the KRL method, standards for screeds and levelling compounds, the introduction of quality levels for the smoothness of subfloors and the future regulation of construction products against the background of the ECJ judgement.

TKB worked with the German Association of Manufacturers of Flexible Floor Coverings (FEB) on the evaluation of floor coverings and adhesives using sensors, the official regulation of construction products and the impact of the introduction of new plasticisers into PVC floor coverings on their bonding properties.

The collaboration with the German Central Association of Parquet and Flooring Installation Companies (ZVPF), which is the most important body for this sector, proved to be very useful, largely due to the involvement of TKB in the ZVPF expert advisory committee. The technical and regulatory issues that affect the trade and the installation materials it uses will be covered regularly in the ZVPF expert conference, as well as in the TKB conference. This includes areas such as the principles of measuring the moisture levels of mineral subfloors. As a result, TKB contributed to ZVPF technical information sheet 1 "Assessing the seams in installed needle felt floor coverings" and ZVPF technical information sheet 2 "Quality requirements for the smoothness of levelled surfaces".

On the important subject of determining the moisture levels in screeds and identifying their readiness for the installation of floor coverings, the ZVPF experts are using the KRL method in parallel with the CM method to measure moisture in support of TKB's efforts to promote the KRL method.

On a European level and in particular as part of the FEICA Working Group Construction, TKB representatives have contributed to the implementation of the Construction Products Regulation, supported the introduction of European standards for installation materials and promoted the use of German sample environmental product declarations (EPDs) in a European context. TKB's involvement in FEICA ensures that it receives information about planned European regulatory measures at an early stage, such as the obligation to provide information to national poison centres, the restrictions on the use of products containing diisocyanates and the Circular Economy Act.

Issues relating to building legislation

TKB is continuing to work towards replacing national regulations with harmonised European standards. The in-depth dialogue with the DIBt about the regulation of sealing compounds is ongoing. The inclusion of emissions requirements in the revision of EN 14293 and EN 14259 is intended to meet Basic Requirement 3 (hygiene and health) of the Construction Products Regulation. This will make it possible to replace the German requirement for a general construction approval with a CE mark.

In October 2015, the European Court of Justice (ECJ) passed its judgement C-100/13, which declared the additional approval requirement for CE-marked products to be invalid. As a consequence of this, the DIBt withdrew its construction products lists. Since early 2016, no construction product approvals have been granted for CE-marked products. The original plan was to introduce a new regulation by 16 October 2016. The model building regulation was amended to allow the existing regulation of building products to be replaced by the regulation of buildings. The necessary notification procedure involving the EU Commission was completed for the accompanying Model Administrative Provisions for Technical Building Rules (MVV TB). Like other affected organisations, TKB believes that the DIBt's approach, which involves circumventing the ECJ judgement, is incompatible with European law. We await further developments.

In addition, TKB is playing an active and critical role in support of the efforts of the German Environment Agency (UBA) to introduce odour testing for building products. In addition to TKB's involvement in the sensors working group of the Committee for Health-Related Evaluation of Building Products (AgBB-AG Sensorik), a study was carried out by the Fraunhofer Institute for Wood Research WKI to determine whether a valid method for identifying odours was already available. This involved analysing all the influencing factors, most importantly to determine whether reproducible results could be achieved with a reasonable amount of work. As things currently stand, no requirements relating to the odour of building products will be incorporated into the DIBt requirements. However, it does seem likely that the criteria for awarding Blue Angel certification will be extended to include a sensor test. This would take effect in 2020 at the earliest.

Hazardous materials/health and safety at work

The GISCODE groups D, S, RU, RS, CP 1 and ZP 1 have been revised. Following the revision of the RE groups in consultation with the German Industry Association for Construction Chemicals (DBC), there will in future only be four RE groups. Two additional GISCODE groups will be created for calcium sulphate levelling compounds that require labelling.

Together with the liability insurance association for the construction industry (BG Bau), investigations have been carried out into the use of silane-based primers to determine whether compliance with occupational exposure limits is needed, in a similar way to GISCODE RS 10 products, or whether another GISCODE group should be introduced.

A ranking system for epoxy resin products is also being developed in collaboration with BG Bau. After this, the aim is to assess the risk potential of the different products on the basis of the substances they contain using a computer model.

On a European level, efforts are being made to tighten up the labelling of products containing methylisothiazolone. Without changes in the formulations, this would result in the requirement to label a large number of dispersion products. A joint position statement has been drawn up on this subject with the DBC and the Association of the German Paint industry and has been submitted to the relevant authorities.

The environment/consumer protection/sustainability
On the basis of the collection of formulations and raw materials compiled by TKB and the DBC, sample EPDs were made available for all the relevant flooring installation materials (dispersion primers and adhesives, single-component PU primers and adhesives, two-component PU adhesives and levelling compounds, SMP adhesives and silane primers, two-component EP adhesives and primers, mineral levelling compounds). FEICA was responsible for transferring the sample German EPDs to a European level and for introducing them throughout Europe. Its efforts in this area also cover the US market. The EPDs will form the basis for the proof of sustainability required as part of the CPR's Basic Requirements for Constructions Works (BRCW) 7.

Technical Committee
DIY and Consumer Adhesives (TKHHB)

The Working Group and Technical Committee DIY and Consumer Adhesives regularly hold joint meetings and monitor a variety of legal activities at a European and national level. The main interest is focused on regulations and topics concerning adhesives in small packages, as far as these are intended for private end users. Important topics in this area include:

- Identification and packaging of adhesives and the obligation to provide information (CLP Regulation)
- Requirements in certain areas where adhesives are applied (appliance and product safety act/toys regulation/medical products act)
- Other normative/legal limitations and requirements regarding adhesives

Three leaflets have been published and made available on IVK's homepage:

1. Labelling guidelines
2. Guidelines on implementing the Toys Directive
3. First aid information relating to the use of instant glues for adhesive bonding

A draft of a new leaflet entitled "Leitfaden zur Abgrenzung von behandelten Waren (Kleb-stoffe/Spachtelmassen/Dichtstoffe) und Biozidprodukten" (Guidelines on keeping treated articles (adhesives/fillers/sealants) and biocidal products separate) has been drawn up and the labelling guidelines have been revised but not yet approved.

Identification and packaging of adhesives and the obligation to provide information (CLP Regulation)
According to Art. 4 of the CLP Regulation, the manufacturer or importer must
- classify substances and preparations prior to marketing them
- package them according to the classification and
- label them.

The technical regulation for hazardous materials, TRGS 200, summarises the applicable rules:
- special labelling regulations for substances and preparations that are available to the general public, 6.7; 10.2
- simplification of labelling requirements and exceptions, 7.1
- implementing labelling regulations, 9

Article 45 of the CLP Regulation specifies that notices must be sent to information and treatment centres for poisoning. The notified bodies receive all information from the importers/manufacturers responsible for marketing products as well as the downstream users of the products to enable them to comply with their tasks. Any subsequent changes to products must be reported as well. The reporting requirements for cleaning agents and solvents (detergent directive) are even more far-reaching.

Requirements in certain areas where adhesives are applied (medical products act/appliance and product safety act/toys regulation)
As a general rule, a product may only be sold if it is designed in such a way that the health and safety of users or third parties are not at risk if used as intended or in a manner that can be foreseen.

This affects the manufacturers of items ready for use. Adhesives are not directly subject to this regulation but are indirectly affected through the end product if the safety (and usability) of an item depends on the suitability of the adhesive.

Adhesives as toys or components in toys for the purpose of the toys regulation/appliance and product safety act – toys directive – EN 71.
EN 71 was revised by order of the EU Commission. It is based on the safety requirements of the Toys Directive 88/378/EEC, whereby toys, i.e. products which are intended for children younger than 14, must be safe prior to being sold. Compliance with the requirements defined in EN 71 is documented by the CE mark. Testing may be made by the manufacturer or by a testing authority; alternatively an EU type approval test may be performed if compliance with EN 71 cannot otherwise be established.

Adhesives as accessories within the meaning of the Medical Devices Act (MPG)
Adhesives for the purpose of MPG can occur as an accessory (as in bandages for instance) and thus are not subject on their own to CE labelling for medical supplies. Nonetheless, manufacturers of medical products may still request compliance with the applicable require-ments. To ensure conformity with the requirements of EU directive 93/42/EEC, it is especially critical to implement Appendix I – Basic requirements.

The diversity and wide variety of medical products which present different levels of complexity and risk mean that manufacturers must apply the "basic requirements" to their specific product. In doing so, "harmonised standards" for the EU directive must be applied, for instance the standard "Biological evaluation of medical devices – Part 1. Evaluation and testing (ISO 10993- 1:2009)." The interests of the medical products industry are safeguarded by BVMed (Tel. +49 (0) 30/2 46 25 50) www.bvmed.de.

Product information in emergencies (DIN EN 15178), labelling checklist
The purpose of the product information standard DIN EN 15178:2007-11 is to improve the identification of products in the event of emergencies. The letter " i " placed near the bar code on packaging refers to the trade or product names or the number under which the product is registered or officially approved.

TRGS 200 "Classification and identification of substances, preparations and products" offers another template for a checklist for labelling issues, which was revised by TKHHB.

Initiatives for Assessing Emissions
The committee for the health assessment of building products (AgBB) has issued a product-related assessment method. VOC emissions are tested on the basis of a test chamber method. In addition, individual substances are measured and assessed, and the LCI (Lowest Concentration of Interest) is applied as the limit value. The latest version of the assessment chart and the appended LCI materials list can be downloaded at https://www.umweltbundesamt.de/themen/gesundheit/kommissionen-arbeitsgruppen/ausschuss-zur-gesundheitlichen-bewertung-von.

At present, only building products are affected by the regulations of Germany's Institute for Building Technology (DIBt). DIY and consumer adhesives are unaffected. The situation in France is different. There, DIY and consumer adhesives are subject to a labelling system based on emissions classification between C and A+. In Belgium, a regulation has been announced that concerns flooring adhesives but not DIY or consumer adhesives.

The EU Commission is giving a high priority to the issue of "indoor air" and has commissioned various projects and studies. Relevant test standards in CEN/TC 351, including prEN 16516, are to be harmonised within the framework of the Building Products Directive (EU/305/2011).

At the request of the Executive Board, GEV has offered to label all kinds of adhesives, including those which are not building adhesives, using the EMICODE labelling system, in as far as its requirements are met.

Packaging Ordinance, recycling bins, circular economy
The German Packaging Ordinance (VerpackV) requires manufacturers to take back packaging. In practice, this is equivalent to obliging them to take part in a waste disposal system, because it is no longer possible for them to take back packaging themselves and participation in industry schemes is accompanied by a strict obligation to provide proof. In addition, there is an obligation to report the typical quantities of packaging for end consumer products to IVK. However, IVK does not publish this information with the quantities.

The objective of the German Recycling Act is to reduce resource use, protect the environment and develop the market for secondary raw materials even further. For this purpose, non-packaging of similar material will be collected for disposal with used packaging. This will be financed by the manufacturers of the products as part of their product responsibility. The municipalities also aim to benefit from this, which is why a political dispute is currently taking place about whether the waste disposal industry or the municipalities will be responsible for organising the scheme.

A working draft of the Act on the further development of the separate collection of reusable household waste is now available. The Federation of German Industries (BDI) in particular believes that the private waste disposal industry is more suited to the task. In a statement issued by the German Chemical Industry Association (VCI), IVK argued that used adhesive tape is not suitable for recycling and therefore should not be included in the cost assessment. Instead it should be disposed of with the residual waste. Adhesives are not directly affected by the regulations.

The Circular Economy Act is also being revised with the aim of increasing the proportion of recycled materials by abandoning the approval of thermal recycling. VCI is opposed to this and its justification is that removal of the fuel value clause (11,000 kJ/kg) would increase the amount of firing using organic energy sources, which makes no environmental sense. Coal has a much lower fuel value of 8000 kJ/kg and therefore larger quantities of it would have to be used than is the case with the plastic waste that is currently approved. In addition, the amount of red tape would increase, staff costs would rise and the waste flows would change without bringing any added value for people or the environment.

Unit pricing/Prepackaging Legislation
In Germany, businesses are obliged to indicate the base prices (i. e., price per kg or litre). This also includes adhesives, unless they are below the minimum limit for base price labelling of 10g or 10 ml (Section 9 Para. 4 of PAngV (German Price Ordinance)). That is why cyanoacrylate adhesives are generally excluded from base price indication.

Section 7 of FPV (German Prepackaging Ordinance) still specifies an obligation to indicate filling quantities by weight. Since viscous adhesives are labelled on the basis of volume in European countries outside Germany, the Working Group for DIY and Consumer Adhesives and the Technical Board has discussed an adaptation within the meaning of harmonisation. The technical switchover also involves disadvantages due to the filling quantities change. The EU Nominal Quantities Directive 80/232 dictates that EU member states are not authorised to prohibit the fill quantities mentioned in the appendices for certain products that are for sale or to limit filling quantities, see Article 5. The German Prepackaging Ordinance (FertigpackungsVF) does not include any sales limitations classified according to filling quantities.

Technical Committee
Wood/Furniture Adhesives (TKH)

In recent years, there has been an increasing focus on health and safety at work, environmental conservation and consumer protection. In these areas, the advisory and supervisory expertise of TKH is in great demand. The committee's contacts with associations and institutes are very important and have been of valuable help over the years.

TKH meetings are regularly held on the premises of a variety of institutes such as the Institute for Wood Technology (IHD) in Dresden. Here TKH was informed about the results of research into the development of a testing procedure for the bonding quality of laminate wood flooring. The results of the two-year research project involving the IHD, the Austrian Forest Products Research Society, the German Association of the Wood Flooring Industry and 28 manufacturers of wood flooring, adhesives and coatings have already been incorporated into European standards and will form the basis for a standardised procedure for assessing the quality of wooden floors in Europe.

The series of data sheets produced by the TKH expert panel, which reflect the best available technology in specific areas of adhesive use, has been constantly added to and updated. The data sheets have now been translated into English and can be downloaded from the IVK website.

In addition to its internal activities, TKH also takes part in working groups covering a number of different industries, such as the profile wrapping working committee and the initiative group on three-dimensional furniture fronts.

Several research projects have taken place over recent years covering the manufacturing processes for three-dimensional furniture fronts. The results of these projects have been integrated into the revised version of the 3D Furniture Front Production Quality Guide, which was completed in May 2017. It is available in German and English at www.klebstoffe.com.

TKH also contributed its expertise to the revision process for one set of VDI (Association of German Engineers) guidelines "VDI RL 3462-3 Reducing emissions – Wood processing and woodworking; Processing and finishing wood materials".

TKH is regularly represented on the CEN TC 193 SC1 Adhesives for Wood standards committee and its subcommittees.

One important development was the publication of the standards DIN EN 204 (Classification of thermoplastic wood adhesives for non-structural applications) and DIN EN 205 (Adhesives. Wood adhesives for non-structural applications. Determination of tensile shear strength of lap joints) as ISO DIS 19209 and ISO DIS 19210.

It is very important that standards and proposed standards are monitored with regard to their practical relevance. Round robin tests within TKH can determine which testing procedures actually yield reproducible results. The effort required to perform the testing can also be evaluated.

Applying the collective expert knowledge of the industry within TKH for the benefit of users is possible only with the commitment of the member companies and their willingness to give their employees the opportunity to carry out work for the association.

Technical Committee
Adhesive Tapes (TKK)

The activities of the Technical Committee Adhesive Tapes (TKK) mainly concern standards issues relating to measurement processes on a European and an international level. However, the specific requirements of a variety of industries, including the automotive industry, also form the starting point for the committee's projects. TKK organises independent activities, but also acts as a catalyst by initiating publicly funded research projects and supporting them in the relevant committees via representatives of its member companies.

The different regional methods for measuring adhesive strength, shear resistance and tensile strength have been harmonised in collaboration with the USA and Japan. Afera (the European Adhesive Tape Association) was responsible for submitting the new harmonised methods to the ISO secretariat for use in countries throughout the world. The harmonisation process has now been completed on a European level (CEN) and an international level (ISO 29862, ISO 29863 and ISO 29864).

Because of a change in the organisational structure of CEN, Afera is no longer a direct member of TC 193/WG7. DIN, which hosts the CEN secretariat on behalf of Afera, will take over responsibility for all the issues relating to the maintenance and revision of the test methods. Lutz Jacob will continue to act as the contact person between DIN and Afera.

As has already been reported, new test methods are being developed by the Global Tape Forum (GTF) which will subsequently be published as GTF test methods. The *Shear Adhesion Failure Test Method* is currently being introduced globally as GTF 6001. At the GTF meeting held in Beijing in May 2016, the Global Test Method Committee decided to adopt the *Thickness and Width and Length* test methods, which had already been harmonised on a worldwide basis, as new GTF test methods 6002 and 6003. In addition, JATMA (the Japanese Adhesive Tape Manufacturers Association) has proposed the development of a new test method for measuring adhesive strength on different steel sheet surfaces.

Afera has carried out several ring tests using the new *Loop Tack Test Method*. The results have been good, but are not yet sufficiently accurate. Improvements to the test method are planned and will be evaluated as part of a global ring test.

Afera will organise the next Global Tape Forum meeting in Munich in June 2018.

The PSTC (Pressure Sensitive Tape Council) has drawn up a proposal for harmonising and introducing new GTF test methods. A vote will be held on this at the global meeting in Munich.

The PSTC has developed a statistical method for determining measurement uncertainties when using test methods. Afera is currently working on adapting it for use in Europe.

Afera has produced a new version of its Test Methods Manual, which contains the latest adaptations and the GTF test methods. The manual will be published before the end of 2017, but will be available only in digital form.

The long-term durability of bonds is particularly important in the automotive industry and also the construction industry. It goes without saying that it must be possible to provide reliable information about the long-term behaviour of bonds created using adhesive tapes. The relevant tests are generally costly and highly time-consuming and, as a result, the development cycles are very long. For this reason, TKK has worked with the Fraunhofer Institute for Manufacturing Technology and Advanced Materials (IFAM) to apply for funding for a research project in this area with the aim of reducing the duration of a 50-day test to just a few days without introducing more radical test conditions. The project was run by IFAM as part of an AiF (German Federation of Industrial Research Associations) research programme. The results showed that physical test methods can highlight changes in PSA (pressure sensitive adhesive) systems caused by thermal or hydrothermal ageing at an early stage, but this does not necessarily make it possible to draw conclusions about the results of a long-term ageing test. One central issue in this context is the applicability of the Arrhenius relationship with regard to the reaction kinetics of PSA systems. It is clear that there is a need for more fundamental research. The requirement has been structured and formulated by TKK, together with IFAM, in order to justify a follow-up to the project described above. The contribution made by TKK included proposals for selecting relevant and representative adhesive tape systems and also suggestions for test scenarios. A project outline prepared by IFAM was discussed by GAK (the joint committee on adhesives technology) and was very well received, in part as a result of the dedicated support given to it by TKK. The full application is now ready and the final evaluation process for the authorisation of public funding will soon be completed. Several member companies of TKK have already expressed their interest in providing practical support for the project in the form of model systems.

TKK is currently considering initiating another research project relating specifically to adhesive tapes. This also concerns the requirements of vehicle manufacturers and, in particular, the public transport sector. It involves the fire protection characteristics of adhesive tapes, including flame retardance and smoke properties. A closer investigation shows that there is relatively little information about the mechanistic relationship between the individual components of the tapes or their interaction with substrates. By contrast a variety of test standards is available, but the situation with regard to adhesive tapes is often unclear. TKK is in the process of formulating a relevant research topic.

A very different aspect of the use of adhesive tapes concerns the issues relating to food legislation. This applies in particular to tapes used in food packaging. During the course of TKK's work, the issues that have arisen primarily concern the evaluation of special applications in the context of food legislation, for example the tapes for splicing film webs that are used for packaging. A more detailed analysis of the issues has shown that the information produced by the Technical Committee Paper and Packaging Adhesives (TKPV) and, in particular, the committee's technical information sheets are often ground-breaking. As a result, TKK has had a representative on TKPV since the beginning of the year and is involved in its work. The results of this work are fed back directly to TKK.

As part of its activities, TKK is continuing to support the European Association of Adhesive Tape Manufacturers (Afera). TKK also contributes reports on conferences and events organised by the adhesive tape industry and on new products to the Afera News newsletter..

Technical Committee
Paper and Packaging Adhesives (TKPV)

Adhesives for products intended to come into contact with food

As in previous years, the Technical Committee Paper and Packaging Adhesives continued to focus on adhesives for food contact materials during the 2015/2016 period. It is still the case that adhesives used for producing materials and articles intended to come into with food are not specifically covered by EU regulations. Due to the fact that adhesives are part of food contact materials, they are, however, subject to assessment under the provisions of food legislation (Framework Regulation (EU) No 1935/2004). If adhesives used for products that come into contact with food are based on materials which are also used to produce plastics for products that come into contact with food, information from the corresponding EU regulations can be applied for a risk assessment based on Article 3 of Commission Regulation (EU) No 1935/2004. In this context, Commission Regulation (EU) No 10/2011 came into force on January 2011 and replaced Directive 2002/72/EC of 6 August 2002. It consolidated all the material lists from the annexes and all the changes and amendments in one regulation. Since then it has been supplemented and revised by seven further regulations.

For substances that are not listed there, the national European regulations, such as the recommendations of Germany's Federal Institute for Risk Assessment (BfR) or the Royal Decree RD 847/2011, still apply. Commission Regulation (EU) No 10/2011 states that adhesives may also be made of substances that are not approved in the EU for the production of plastics.

Commission Regulation (EC) No. 2023/2006 on good manufacturing practices for materials and articles specifies the principles of "Good Manufacturing Practices - GMP", as required by Article 3 of Commission Regulation (EU) No 1935/2004 (EU Framework Regulation for Materials and Articles Intended to Come into Contact with Food). This regulation applies to all articles and materials listed in Annex 1, including adhesives. Strictly speaking, the GMP regulation does not apply to raw materials that are used in the corresponding adhesives. Nonetheless, the materials do have to fulfil specifications which enable adhesive manufacturers to work in accordance with GMP. Guidelines entitled "Good Manufacturing Practices" are available to help adhesives manufacturers to comply with the regulation.

The Union Guidance on Regulation (EU) No 10/2011 was published on 28 November 2013 by the services of the Directorate-General for Health and Consumers and is available in English only.

It is intended to help with interpreting and implementing questions about conformity decla-rations, conformity activities and providing information along the supply chain for food contact materials.

In this EU guidance document, adhesives for the manufacture of food contact materials made of plastic or plastic composites are described as "non-plastic intermediate materials". Section 4.3.2 lists all the relevant information that a manufacturer of "non-plastic intermediate materials" should provide throughout the supply chain.

TKPV technical information sheets 1 to 4 have been revised on this basis. The information sheets already cover all aspects of the Union Guidance on Regulation (EU) No 10/2011 with regard to adhesives. Nevertheless, a formal revision taking into account the Union Guidance and all the new regulations and directives was unavoidable. German and English versions of each infor-mation sheet are available on the association's website.

The subject of mineral oils in foods is currently occupying German legislators, who would like to introduce new regulations for food packaging made of recycled cardboard. The main source of these oils has been identified as the mineral oil-based printing inks from newspaper printing. However, in Regulation No 10/2011/EC, materials that contain mineral oil saturated hydro-carbons (MOSH) and/or mineral oil aromatic hydrocarbons (MOAH) are listed and approved for the production of food contact articles made from plastic. Examples are medical white mineral oils, which are frequently used in cosmetics, and liquid paraffin, which is used in organic farming. TKPV has drawn up an internal paper for the members of IVK which provides information on possible sources of MOSH and MOAH.

The consultation process for the fourth draft of the "Mineral oil ordinance" produced by the German Federal Ministry of Food and Agriculture is currently underway. This states that food contact materials made from paper, paperboard or cardboard can only be manufactured from recycled fibre and distributed if they have a functional barrier which ensures that no MOAH can be transferred from the food contact materials into the food. A transfer is considered not to have taken place up to a detection limit of 0.5 milligrams of all MOAH per kilogram of food or food simulant.

The limits originally proposed for mineral oil aliphatic hydrocarbons were removed, together with the concentration limits in cardboard. A barrier is not needed if manufacturers or distributors of food contact materials have taken other appropriate measures to prevent MOAH from being transferred into the food. It is not yet clear whether Germany will take independent action in this area or whether the final draft will be handed over to the EU Commission.

TKPV has produced the final version of a technical information sheet that provides information about mineral oil hydrocarbons in adhesives used in cardboard packaging. This will soon be available on the IVK website.

Adhesives in Paper Recycling:
Another important and ongoing aspect of TKPV's work involves assessing the impact of adhesive applications on paper recycling. In connection with the voluntary commitment made by the graphic industry, the EU Commission and Germany's Federal Environment Ministry have called for higher recycling quotas for paper.

In this context, TKPV is taking part in a detailed dialogue with the paper industry and scientific institutions. During the course of a number of scientific events on this topic, TKPV has presented its views on the problem of adhesive contaminants.

With the support of TKPV, the International Research Group on De-Inking Technology INGEDE has developed the testing procedure "INGEDE Method 12 - Assessment of the Recyclability of Printed Paper Products - Testing of the Fragmentation Behaviour of Adhesive Applications". This testing procedure determines whether an adhesive can be removed using a screening process. Criterion 3 of the Commission Decision of 16 August 2012 establishing the ecological criteria for the award of the EU Ecolabel for printed paper (2012/481/EU) specifies that all adhesives must be tested according to the INGEDE Method 12 or equivalent. Only those adhesives that have been proven to be removable are used in printed products. The USER'S MANUAL FOR THE APPLICATION FOR PRINTED PAPER (Commission Decision 2012/481/EU), which is part of the Commission Decision, states that water-based adhesives do not need to be tested. According to the Commission Decision of 2 May 2014 establishing the ecological criteria for the award of the EU Ecolabel for converted paper products (2014/256/EU), it is now also possible to apply for the EU Eco-label for envelopes and paper bags. For adhesives, Criterion 4 – Recyclability (b) is relevant. Non-soluble adhesives may be used only if they can be proven to be removable. No definition of the term "non-soluble" is provided.

TKPV is now working on "clustering" hot melt adhesives. The aim of this project is to identify existing physical measuring methods that can be used as a substitute for the INGEDE Method 12. RAL and INGEDE are supporting the project. The test series have been completed and the results evaluated. A substitute evaluation is now available in the form of a minimum softening point and a minimum thickness. The final meetings with INGEDE have not yet taken place. Afera (the European Adhesive Tape Association) has arranged for TKPV to be represented on the ERPC (the European Recovered Paper Council).

TKPV supported INGEDE project 147 15 relating to the behaviour of water-dispersible adhesives during the recycling process. The project has been completed, but was then rejected because of a score card. A new approach is currently being developed which is intended to lead to another research project concerning the evaluation of water-dispersible adhesives.

REACH:
Another area of emphasis of TKPV's work included and still includes activities relating to the new European Chemicals Regulation "REACH" (Registration, Evaluation, Authorisation and Restriction of Chemicals). Under the provisions of REACH, it is necessary to provide substance data and exposure information to ensure that substances are handled safely. For this reason, downstream users of substances must inform the registrants about their use. In order to guarantee that this

communication takes place, ECHA has developed a "Use Descriptor" model. On the basis of this model, TKPV has developed use scenarios for the production and the common applications of paper and packaging adhesives. The use scenarios have been included in the FEICA use-mapping for adhesive production and application.

The second ATP of the CLP Regulation specified new rules for materials in mixtures that may cause an allergic reaction. If there is a labelling limit for a material, the mixture is labelled as follows from one tenth of this concentration: "EUH208 contains (name of the sensitising substance). May produce an allergic reaction". This applies primarily to water-based dispersion adhesives that contain preserving agents such as isothiazolinone. The new labelling requirements in Article 58(3) of the Biocidal Products Regulation 528/2012 must be complied with in future, in particular in the case of in-can preservatives for water-based dispersion adhesives.

Changes to the labelling regulations for methylisothiazolinone are likely to be introduced. These will be adapted to correspond with the labelling regulations for chloromethylisothiazolinone/ methylisothiazolinone (3:1). As products that require no labelling will then fall below the limit for methylisothiazolinone, the industry is currently looking for alternatives.

Standards activities:
TKPV has been successful in ensuring that German manufacturers of paper and packaging adhesives have expert representation on a range of committees and working groups, such as the "Adhesives for Paper, Cardboard, Packaging and Disposable Sanitary Products" work group of the European standardisation committee CEN/TC 193 and the German mirror committee DIN/NMP 458.

TKPV supported the working group ISO/TC 34/SC 17/WG 5, which developed the technical specification ISO/TS 22002-4 "Prerequisite programmes on food safety – Part 4: Food packaging manufacturing". With the aid of this technical specification, packaging manufacturers and suppliers can ensure that their production processes comply with GMP and have them certified. A German translation of the specification is already available. A GFSI ranking has not yet been assigned.

Networking:
IVK has taken out a subscription with the Federal Institute for Risk Assessment (BfR) on behalf of TKPV to ensure that TKPV is informed directly about changes to the recommendations for food contact materials. In addition, at TKPV's request IVK is now a member of the German Federation for Food Law and Food Science, which gives it rapid access to a wide range of information sources.

FEICA:
Due to their importance at a European level, the topics of adhesives for materials and articles intended to come into contact with food, MIGRESIVES, FACET, GMP and REACH continue, at the request of TKPV, to be discussed extensively with the European Paper and Packaging and FACET working groups as well as the REACH working group at FEICA. The aim is to establish a joint position for the European adhesives industry and to find solutions to these important issues. TWG PP has also investigated mineral oils in foods and, like TKPV, has produced a technical information sheet on the subject.

Other topics dealt with:
- The work of the Postpress Technical Committee of the Research Institute for Media Technologies (FOGRA)
- The research forum of the PTS
- The work of the Industry Association for Food Technology and Packaging (IVLV) in relation to food packaging
- Cosmetics packaging

Technical Committee
Footwear and Leather Adhesives (TKS)

The Technical Committee Shoe Adhesives (TKS) coordinates public relations activities on behalf of German manufacturers of adhesives for footwear materials. In addition to that, the committee supports German and international standardisation activities activities and acts as the point of contact for market segment-specific technical evaluations and information specific to certain market segments within as part of the work of thethe activities of German Adhesives Association (IVK) and the German Chemical Industry Association (VCI) as well as for regulatory affairswith regard to regulatory issues.

Standardisation
The committee focusesis focused on cooperating insupporting the development of German and European standards to define the basic properties of footwear adhesives.

These activities include, for example, standards for:
- Minimum requirements on for footwear bonds (requirements and materials)
- Tests to check bond strength in footwear (peel resistance tests)
- Processing (determining optimum the ideal activation conditions and sole positioning tack)
- Durability (colour change caused by migration, thermal resistance of lasting adhesives)

The provision and availability of standardised reference test materials and reference test adhesives are crucial for the ability to perform a large number of tests. The sSelection and specification of the reference test materials and adhesives are subject to continuous testing and updatingcontinuously evaluated and updated on the basis ofin accordance with the current state oflatest technology. TKS is currently preparing for an audit in this area in 2018.

The TKS leaflet, which isinformation sheet entitled "Trouble Shooting bei der Schuherstellung" ("Troubleshooting in fFootwear Mmanufacture)" and is available for download atto download from IVK's website. It, provides assistance and tips for emergencies and service issues that occurring during the use of adhesives.

Training
Providing training in adhesive bonding for employees in the footwear industry is another of the tasks that TKS is responsible for. The first courses took place in Zweibrücken in 1990 and in Pirmasens in 1992.

TKS aims to share adhesives knowledge that makes it possible to identify practical problems more quickly, develop solutions and take measures to minimise and even eliminate potential sources of faults in the future.

Against this background, TKS has developed a training concept in conjunction with IFAM/Bremen and PFI/Pirmasens. It takes the form of a practical seminar called "Angewandte Klebtechnik in der Schuhindustrie" (Applied adhesive technology in the footwear industry). It was held for the first time in November 2005 with more than 20 participants. They gained an in-depth understanding of adhesive technology and the associated issues. The training focuses on providing practical examples to demonstrate best practices in bonding and typical problems. The participants can apply what they have learned during the course of practical exercises, which makes the theoretical background information easier to understand for users working in shoe production.

Technical Committee
Structural Adhesives and Sealants (TKSKD)

In light of the growing number of adhesive applications in which adhesive bonds play a structural role, the association's newest body, the Technical Committee Structural Adhesives and Sealants (TKSKD), has been focusing on a range of current technical issues relating to structural adhesives and sealants for the last 15 years and supports the manufacturers of structural adhesives and sealants which are members of the corresponding work group (AKSKD), together with raw material producers and research institutes involved in this field.

Because structural adhesives are used in many different industries (for example, car, train and aircraft production and shipbuilding, the electrical and electronics industry, the household appliances industry, medical technology, the optical industry, mechanical engineering, plant and equipment engineering, wind power and solar energy) for a variety of applications, TKSKD covers a wide variety of technical topics.

The committee's work includes:
• Training in adhesive technology: Thorough training in adhesive technology and a good understanding of its applications help to ensure that adhesives are used successfully, correctly and in accordance with the requirements of specific production processes and are therefore in the interests of adhesive manufacturers. This is reinforced by the DIN 2304-1 standard, which was published in early 2016. It requires employees who take responsibility for planning and implementing safety-related structural bonds to have the appropriate qualifications. The

committee supports the three-stage DVS/EWF training concept, which has also gained international acceptance, and the preparation of the guidelines "Kleben – aber richtig" ("Adhesive Bonding – the Right Way"), which are available on the IVK website.

- Promoting research: TKSKD also sees itself as a bridge between industry and scientific activities that take place outside the industrial world and regularly provides information about topics relevant to structural adhesives as part of pre-competitive adhesive research. For instance, the project run by Braunschweig University of Technology to verify the validity of OIT measurements concerning the thermo-oxidative stability of reactive adhesive systems as a cost-effective and time-saving method for improving the temperature stability of adhesive formulations was supported by individual member companies and the latest status of the project has been reported in AKSKD.

In addition, there have been regular reports on the progress of a project supported by the German Federal Ministry of Education and Research (BMBF) concerning the laser pre-treatment of fibre composites and the development of an in-line monitoring procedure for the pre-treatment process.

- Standards activities: The members of the structural adhesives standardisation committee financed by IVK report on a regular basis on standardisation activities at a German, European and international level. The topics covered over the past few years include:

  - Information on the current status of the internationalisation of Germany's rail vehicle standard DIN 6701 "Adhesive bonding of railway vehicles and parts". This standard defines the mandatory requirements placed on companies that manufacture bonded rail vehicles or parts for use on Germany's railways. It is now being turned into a European standard.
  - Information on the work of the standards committee "Quality assurance of adhesive bonds". TKSKD members on this committee have been actively involved in drawing up DIN standard 2304 (Adhesive bonding technology – Quality requirements for adhesive bonding processes – Part 1: Adhesive bonding process chain). They are currently working on supplementary regulations covering the bonding of fibre composites, for example, and they keep the AKSKD members regularly informed about the status of their work.

- Chemicals legislation: As in the previous reporting period, REACH and its requirements for the manufacturers of adhesives and their customers remained a key topic. The member companies are kept regularly informed about current issues. Examples of these include:

  - New inclusions in the list of Substances of Very High Concern (SVHCs).
  - The change in the labelling requirements for methylene diphenyl diisocyanate (MDI) and the current status of the restriction dossier issued by the German Federal Institute for Occupational Safety and Health (BAuA) for diisocyanates in general. The goal is to prevent the introduction of the more far-reaching restrictions on the monomeric diisocyanates that have been requested by some EU member states.
  - The classification of mixtures with organostannic compounds as hazardous materials.
  - The proposals for supplementary classification as toxic for reproduction.

- Testing method for the drinking water approval of anaerobic adhesives: As the guidelines published by the UBA have been withdrawn, there is currently no suitable testing method available for obtaining the necessary drinking water approval for anaerobic adhesives, which are used, among other things, as thread sealants in drinking water systems. An ad-hoc working group within TKSKD has developed practical, standardised guidelines in cooperation with testing institutes and the Federal Environmental Agency (UBA) which permit the use of anaerobic adhesives in drinking water systems.

On the initiative of Henkel and Sika Automotive, which are members of the committee, a joint position statement has been drawn up on the subject of the REACH classification of baffles as a product. Baffles are used in lightweight automotive components to reinforce structural parts. They consist of a combination of a structural part and a material that expands when exposed to a high temperature. The shape of the baffle corresponds to the outline of the structural part and it is incorporated into the structural part in the body-in-white. In the paint curing oven, the material expands and bonds with the structural part. As the shape of a baffle largely determines its functionality, according to REACH it can be categorised as a product. This means that it does not have to be classified on the basis of the CRP Regulation and that no safety data sheets are required.

Public Relations Advisory Board (BeifÖ)

The key task of the Public Relations Advisory Board, which is led by Ulrich Lipper, is to present the German Adhesives Association (IVK) and the key technology of adhesive bonding in all its depth and complexity to the general public in a positive light.

The ongoing public relations work of the German Adhesives Association has proved to be a great success. The subject of adhesive bonding is now firmly established in the print and online media, and the media response is correspondingly high. IVK's press and public relations work generates an average annual circulation of around 150 million copies.

The online press platform www.klebstoff-presse.com, which was designed to meet the special requirements of journalists, and IVK's own internet portal www.klebstoffe.com had a consistently high number of page views. Companies, interested end users and journalists can identify the content that is relevant to them at a glance. The association also has a presence on the main social media platforms, such as Facebook, Twitter and YouTube, with "bonding" now taking place on all channels.

The e-paper "Berufsbilder" (Job profiles) has also led to increased web traffic. It highlights for school and university students the varied career opportunities available in the German adhesives industry. Jobs in scientific, technical and commercial fields are explained in a clear, practical and inspirational way. The objective is to encourage the adhesives experts of tomorrow to become familiar today with the industry and the variety of options it has to offer.

The magazine "Kleben fürs Leben" (Bonding for Life) has become a permanent feature of the association's PR activities. Published once a year, it forms an essential component of the communication strategy of the German adhesives industry and plays a key role in promoting the industry's activities. The magazine is free from any form of product or corporate advertising and focuses on further strengthening the positive image of the adhesives industry and documenting useful characteristics of unique and multifaceted bonding technologies. "Kleben fürs Leben" will be published in 2017 for the ninth time. Since 2013, readers have also been able to access the magazine online as an interactive e-paper. The magazine is distributed in printed form or as an e-paper to the editorial departments of numerous newspapers and TV and radio stations and to important customer organisations.

The IVK promotional video "Faszination Kleben" (The Fascination of Bonding) demonstrates that adhesives are an essential feature of everyday life in the home, in the construction trades and in industry. It also explains why many technologies of the future and current production processes for everyday objects are only made possible by the use of adhesives. In the space of five minutes, viewers not only discover important facts about the chemical aspects of adhesives and how they function, but also find out about the different areas where adhesives are successfully used. Almost every industry, from the automotive sector to electrical engineering, textiles and clothing, relies on bonding to improve the quality of its products and introduce innovations. The video is available in German and in English on the IVK website at www.klebstoffe.com where it can be viewed or downloaded. It can also be found on the IVK YouTube channel "klebstoffe" (adhesives).

Against the background of the German Federal Government's digitisation strategy, the association has worked together with the German Chemical Industry Fund to develop digital educational materials. A range of interactive infographics designed specifically for use on whiteboards, PCs and tablets in schools has been created to explain the subject of adhesives to school students using modern teaching methods. In addition, the association has commissioned the Institute of the Didactics of Chemistry at the University of Frankfurt to develop a teacher training programme on the subject of "Adhesives and bonding in teaching" in order to give a more in-depth insight into the chemistry of adhesives and to highlight links with the school curriculum.

IVK invites representatives from the business and industry press and daily newspapers to attend its annual press event, which highlights the economic growth of Germany's adhesives industry.

Management

The employees of the German Adhesives Association's office take care of coordinating, handling and following up on the wide variety of tasks which come from various committees. The office keeps members up to date on new topics that are important for the industry and its specialised committees. It also serves as an information exchange for the association's members with regard to technical and relevant legal information pertaining to health and safety at work, environmental and consumer protection and competition, chemicals, environmental and food legislation.

The management of IVK perceives itself as a representative and competent partner to the adhesives industry. Within this role, the management represents the technical and economic interests of the industry by maintaining proactive and responsive dialogue with German, European and international authorities, customer and consumer associations, system partners, institutions and the general public. Through active involvement in the advisory boards of major trade shows and trade journals, as well as working committees of various German and European ministries, the management keeps track of and supports adhesives-related topics and projects on all levels and throughout the entire adhesives value chain with its expertise and know-how.

This also applies to research. In its role as a board member of the DECHEMA's (Society for Chemical Engineering and Biotechnology) "adhesives technology" division and its joint committee on adhesive bonding (GAK), the management helps with coordinating publicly funded scientific research projects in the field of adhesives and adhesives technology and contributes to an industry-wide dialogue.

Through a broad portfolio of publications, discussions with interested groups, lectures and teaching assignments, the management effectively communicates the adhesives industry's broad range of services and high potential for innovation as well as the exemplary commitment of its members to protecting workers, consumers and the environment.

PERSONAL DATA

Honorary Membership & Honorary Presidency

During the 2012 annual general meeting, Arnd Picker was made an Honorary Member and at the same time appointed Honorary President of the German Adhesive Association. By doing so, the association and its members recognised Arnd Picker's achievements in successfully positioning IVK during his 16-year tenure of office as President of its Executive Board. The German Adhesives Association is the world's largest and leading association dedicated to adhesive bonding technology and with an extensive service portfolio for its members.

Honorary Members
Arnd Picker, Honorary President
Dr. Johannes Dahs
Dr. Hannes Frank
Dr. Rainer Vogel
Heinz Zoller

Achievement Medal of the German Adhesives Industry

The German Adhesives Association awards the Achievement Medal of the German Adhesives Industry to individuals for their outstanding service to the adhesives industry and adhesives technology.

The medal was awarded to

Marlene Doobe – June 2017
for her decades of commitment to the German adhesives industry. As editor-in-chief of the magazine adhäsion KLEBEN + DICHTEN, she has successfully led this important trade magazine for the adhesives industry for more than 20 years, with professionalism, the highest level of personal dedication and a great deal of passion. During this time and in close cooperation with the German Adhesives Association, she has developed this magazine into an interdisciplinary 360° communication platform for adhesives technology. Marlene Doobe has made an extremely valuable contribution towards promoting close cooperation between research and industry in the field of adhesives technology and has advanced and shaped the adhesives industry in decisive dimensions.

Dr. Manfred Dollhausen – May 2010
in recognition of his successful involvement in standardisation, with which he made a major contribution to documenting and representing adhesive technology "made in Germany" in national, European and international standards. In addition to that, Dr. Dollhausen recognised early on the invaluable significance of technical cooperation between the adhesives industry and the raw material industry in the 1960s, actively forcing cooperation and thus laying the foundation for the system partnership of both industries that is still successful to this day.

Dr. Hannes Frank – September 2007
in recognition of his many years of dedicated service to the German adhesives industry. As a member of the Technical Committee, he promoted and helped shape both adhesive technology and the image of the adhesives industry. That includes in particular his commitment to small and medium-sized enterprises and their potential to innovate, which is essential for technical and economic development. Dr. Frank is also regarded as a successful pioneer in the field of polyurethane adhesive technology. Furthermore, he has promoted an industry-wide strategy on communication and training, and thus helped to establish adhesives as a key technology of the 21st century.

Prof. Dr. Otto-Diedrich Hennemann – May 2007
in recognition of his scientific work, with which he is largely responsible for promoting and shaping the "bonding system". This includes his research in the durability of bonds and the implementation of appropriate simulation processes in the automotive and aerospace industries. His approach to research was to always focus on concrete applications and the development of added value for system partners.

Committees of the
German Adhesives Association (IVK)

Executive Board

Chairman: Dr. Boris Tasche	Henkel AG & Co. KGaA D-40191 Düsseldorf
Deputy Chairman: Dr. Joachim Schulz	EUKALIN Spezial-Klebstoff Fabrik GmbH D-52249 Eschweiler

Additional Members:

Stephan Frischmuth	tesa SE D-22848 Norderstedt
Ansgar van Halteren	Industrieverband Klebstoffe e. V. D-40219 Düsseldorf
Dr. Achim Hübener	Kleiberit Klebstoffe Klebchemie M. G. Becker GmbH & Co. KG D-76356 Weingarten
Patrick Kivits	H.B. Fuller Europe GmbH CH-8001 Zürich
Klaus Kullmann	Jowat SE D-32758 Detmold
Olaf Memmen	Bostik GmbH D-33829 Borgholzhausen
Dr. Bernhard Momper	Celanese Services Germany GmbH D-65844 Sulzbach (Taunus)
Torsten Nitzsche	Sika Automotive GmbH D-22525 Hamburg
Dr. Rüdiger Oberste-Padtberg	ARDEX GmbH D-58453 Witten
Dr. Thomas Pfeiffer	Türmerleim GmbH D-67061 Ludwigshafen

Peter Rambusch	certoplast Technische Klebebänder GmbH D-42285 Wuppertal
Dr. René Rambusch	certoplast Technische Klebebänder GmbH D-42285 Wuppertal
Dr. Rainer Schönfeld	Henkel AG & Co. KGaA D-40191 Düsseldorf
Dr. H. Werner Utz	UZIN UTZ Aktiengesellschaft D-89079 Ulm

Technical Board

Chairman: Dr. Rainer Schönfeld	Henkel AG & Co. KGaA D-40191 Düsseldorf

Additional Members:

Dr. Norbert Arnold	UZIN UTZ Aktiengesellschaft D-89079 Ulm
Dr. Knut Göke	Kömmerling Chemische Fabrik GmbH D-66954 Pirmasens
Prof. Dr. Andreas Groß	IFAM Fraunhofer-Institut für Fertigungstechnik und Angewandte Materialforschung D-28359 Bremen
Daniela Hardt	Celanese Services Germany GmbH D-65843 Sulzbach (Taunus)
Dr. Achim Hübener	Kleiberit Klebstoffe Klebchemie M.G. Becker GmbH & Co. KG D-76356 Weingarten
Dr. Georg Kinzelmann	Henkel AG & Co. KGaA D-40191 Düsseldorf
Christoph Küsters	3M Deutschland GmbH D-41453 Neuss

Dr. Dirk Lamm	tesa SE D-22848 Norderstedt
Dr. Annett Linemann	H.B. Fuller Deutschland GmbH D-21335 Lüneburg
Dr. Hartwig Lohse	Klebtechnik Dr. Hartwig Lohse e. K. D-25597 Breitenberg
Dr. Michael Nitsche	Bostik GmbH D-33829 Borgholzhausen
Matthias Pfeiffer	Türmerleim GmbH D-67061 Ludwigshafen
Arno Prumbach	EUKALIN Spezial-Klebstoff Fabrik GmbH D-52249 Eschweiler
Dr. Karsten Seitz	tesa SE D-22848 Norderstedt
Dr. Christian Terfloth	Jowat SE D-32758 Detmold
Dr. Christoph Thiebes	Covestro Deutschland AG D-51373 Leverkusen
Dr. Axel Weiss	BASF SE D-67056 Ludwigshafen

Technical Committee Building Adhesives

Chairman: Dr. Norbert Arnold	UZIN UTZ Aktiengesellschaft D-89079 Ulm

Additional Members:

Dr. Thomas Brokamp	Bona GmbH Deutschland D-65549 Limburg

Manfred Friedrich	Sika Deutschland GmbH D-48720 Rosendahl
Dr. Frank Gahlmann	Stauf Klebstoffwerk GmbH D-57234 Wilnsdorf
Jürgen Gehring	Bostik GmbH D-33829 Borgholzhausen
Holger Hartmann	Celanese Services Germany GmbH D-65843 Sulzbach (Taunus)
Dr. Hardy Herold	Wacker Chemie AG D-84489 Burghausen
Dr. Matthias Hirsch	Kiesel Bauchemie GmbH u. Co. KG D-73730 Esslingen
Michael Illing	Forbo Eurocol Deutschland GmbH D-99091 Erfurt
Dr. Maximilian Rüllmann	BASF SE D-67056 Ludwigshafen
Helmut Schäfer	Sopro Bauchemie GmbH D-65203 Wiesbaden
Dr. Martin Schäfer	Wakol GmbH D-66954 Pirmasens
Dr. Jörg Sieksmeier	ARDEX GmbH D-58453 Witten

| Hartmut Urbath | PCI Augsburg GmbH
D-40589 Düsseldorf |
| Dr. Steffen Wunderlich | Kleiberit Klebstoffe
Klebchemie M. G. Becker GmbH & Co. KG
D-76356 Weingarten |

Technical Committee Wood and Furniture Adhesives

| Chairman:
Daniela Hardt | Celanese Services Germany GmbH
D-65843 Sulzbach (Taunus) |

Additional Members:

Wolfgang Arndt	Covestro Deutschland AG D-51373 Leverkusen
Holger Brandt	Follmann GmbH & Co. KG D-32423 Minden
Christoph Funke	Jowat SE D-32758 Detmold
Oliver Hartz	BASF SE D-67056 Ludwigshafen
Dr. Thomas Kotre	Planatol GmbH 83101 Rohrdorf-Thansau
Jürgen Lotz	Henkel AG & Co. KGaA Standort Bopfingen D-73442 Bopfingen
Dr. Marcel Ruppert	Wacker Chemie AG D-84489 Burghausen
Dipl.-Ing. Martin Sauerland	H.B. Fuller Deutschland GmbH D-31582 Nienburg
Holger Scherrenbacher	Kleiberit Klebstoffe Klebchemie M. G. Becker GmbH & Co. KG D-76356 Weingarten

Technical Committee DIY and Consumer Adhesives

| Chairman: | tesa SE |
| Dr. Dirk Lamm | 22848 Norderstedt |

Additional Members:

Frank Avemaria	3M Deutschland GmbH D-41453 Neuss
Dr. Nils Hellwig	Henkel AG & Co. KGaA D-40191 Düsseldorf
Dr. Florian Kopp	RUDERER KLEBETECHNIK GMBH D-85604 Zorneding
Aurelia Liar/Dr. Christian Hanf	UHU GmbH & Co. KG D-77815 Bühl
Ulrich Lipper	Cyberbond Europe GmbH D-31515 Wunstorf

Technical Committee Adhesive Tapes

| Chairman: | tesa SE |
| Dr. Karsten Seitz | D-22848 Norderstedt |

Additional Members:

Dr. Thomas Christ	BASF SE D-67056 Ludwigshafen
Thorsten Gurke	Kraton Polymers GmbH D-60327 Frankfurt
Dr. Thomas Hanhörster	Sika Automotive GmbH D-22525 Hamburg
Prof. Dr. Andreas Hartwig	IFAM Fraunhofer-Institut für Fertigungstechnik und Angewandte Materialforschung D-28359 Bremen
Lutz Jacob	RJ Consulting D-87527 Altstaedten

Melanie Lack	H.B. Fuller Deutschland GmbH D-21335 Lüneburg
Dr. Thorsten Meier	certoplast Technische Klebebänder GmbH D-42285 Wuppertal
Jürgen Peters	3M Deutschland GmbH D-41453 Neuss
Dr. Ralf Rönisch	COROPLAST Fritz Müller GmbH & Co. KG D-42279 Wuppertal
Dr. Jürgen K. L. Schneider	TSRC (Lux.) Corporation S.a.r.l. L-1930 Luxemburg
Michael Schürmann	Henkel AG & Co. KGaA D-40191 Düsseldorf
Dr. Miriam Verbruggen	Lohmann GmbH & Co. KG D-56567 Neuwied

Technical Committee Paper and Packaging Adhesives

| Chairman:
Arno Prumbach | EUKALIN Spezial-Klebstoff Fabrik GmbH
D-52249 Eschweiler |

Additional Members:

Kai Biedebach	Bostik GmbH D-33829 Borgholzhausen
Dr. Gerhard Kögler	Wacker Chemie AG D-84489 Burghausen
Dr. Anja Köth	tesa SE 22848 Norderstedt
Dr. Thomas Kotre	Planatol GmbH D-83101 Rohrdorf-Thansau

Dr. Bernhard Momper	Celanese Services Germany GmbH D-65843 Sulzbach
Janet Pohl	Klebstoffwerke COLLODIN GmbH D-60386 Frankfurt
Dr. Werner Praß	Türmerleim GmbH D-67061 Ludwigshafen
Dr. Peter Preishuber-Pfluegl	BASF SE D-67056 Ludwigshafen
Dr. Eckhard Pürkner	Henkel AG & Co. KGaA D-40191 Düsseldorf
Alexandra Roß	H.B. Fuller Deutschland GmbH D-21335 Lüneburg
Dr. Christian Schmidt	Jowat SE D-32758 Detmold
Julia Szincsak	Follmann Chemie GmbH D-32423 Minden

Technical Committee Footwear and Leather Adhesives

Chairman: Dr. Knut Göke	Kömmerling Chemische Fabrik GmbH D-66954 Pirmasens

Additional Members:

Wolfgang Arndt	Covestro Deutschland AG D-51373 Leverkusen
Dr. Rainer Buchholz	RENIA Ges. mbH chemische Fabrik D-51109 Köln
Andreas Ecker	HB FULLER Austria GmbH A-4600 Wels

Technical Committee Structural Adhesives and Sealants

Chairman: Dr. Hartwig Lohse	Klebtechnik Dr. Hartwig Lohse e. K. D-25597 Breitenberg

Additional Members:

Dr. Beate Baumbach	Covestro Deutschland AG D-51373 Leverkusen
Ralf Fuhrmann	Kömmerling Chemische Fabrik GmbH D-66954 Pirmasens
Dr. Oliver Glosch	Weiss Chemie + Technik GmbH & Co. KG D-35708 Haiger
Dr. Stefan Kreiling	Henkel AG & Co. KGaA Standort Heidelberg D-69112 Heidelberg
Dr. Erik Meiß	IFAM Fraunhofer-Institut für Fertigungstechnik und Angewandte Materialforschung D-28359 Bremen
Dr. Karl Michael Müller	Bostik GmbH D-33829 Borgholzhausen
Frank Steegmanns	Stockmeier Urethanes GmbH & Co. KG D-32657 Lemgo
Julius Weirauch	3M Deutschland GmbH D-41453 Neuss
Artur Zanotti	Sika Deutschland GmbH D-72574 Bad Urach

Public Relations Advisory Board

Convenor: Ulrich Lipper	Cyberbond Europe GmbH D-31515 Wunstorf

Additional Members:

Rolf J. Blaas	Dow Deutschland Anlagengesellschaft mbH D-65824 Schwalbach
Holger Elfes	Henkel AG & Co. KGaA D-40191 Düsseldorf
Ansgar van Halteren	Industrieverband Klebstoffe e. V. D-40219 Düsseldorf
Oliver Jüntgen	Henkel AG & Co. KGaA D-40191 Düsseldorf
Timm Koepchen	EUKALIN Spezial-Klebstoff Fabrik GmbH D-52249 Eschweiler
Thorsten Krimphove	WEICON GmbH & Co. KG D-48157 Münster
Dr. Christine Wagner	Wacker Chemie AG D-84489 Burghausen

Working Group Building Adhesives

Chairman: ARDEX Gmbh
Dr. Rüdiger Oberste-Padtberg   D-58453 Witten

Working Group Wood and Furniture Adhesives

Chairman: Jowat SE
Klaus Kullmann   D-32758 Detmold

Working Group Industrial Adhesives

Chairman: Henkel AG & Co. KGaA
Dr. Boris Tasche   D-40191 Düsseldorf

Working Group Adhesive Tapes

Chairman: certoplast Technische Klebebänder GmbH
Peter Rambusch   D-42285 Wuppertal

Working Group Paper and Packaging Adhesives

Chairman: Türmerleim GmbH
Dr. Thomas Pfeiffer D-67061 Ludwigshafen

Working Group Raw Materials

Chairman: Celanese Services Germany GmbH
Dr. Bernhard Momper D-65843 Sulzbach (Taunus)

Working Group Structural Adhesives and Sealants

Chairman: Sika Automotive GmbH
Torsten Nitzsche   D-22525 Hamburg

Working Group DIY and Consumer Adhesives

Chairman: tesa SE
Dr. Dirk Lamm   D-22848 Norderstedt

Working Group Foam Adhesives

Chairman: Jowat SE
Falk Potthast   D-32758 Detmold

Working Group Footwear and Leather Adhesives

Chairman: Kömmerling Chemische Fabrik GmbH
Dr. Knut Göke   D-66954 Pirmasens

Management

Ansgar van Halteren	Senior Executive
Dr. Axel Heßland	Managing Director, "Technology & Environment"
Klaus Winkels	Managing Director , "Law"
Michaela Szkudlarek	Assistant to Senior Executive
Danuta Dworaczek	"Technology & Environment"
Martina Weinberg	Conferences and Conventions
Natascha Zapolowski	"Technology & Environment"
Elke Wegerich	"Internet, Social Media, Communication"

Honorary President

Arnd Picker   Rommerskirchen

Honorary Members

Dr. Johannes Dahs	Königswinter
Dr. Hannes Frank	Detmold
Arnd Picker	Rommerskirchen
Dr. Rainer Vogel	Langenfeld
Dipl.-Chem. Heinz Zoller	Pirmasens

Holders of the Medal of the German Adhesives Industry

Marlene Doobe	Eltville
Dr. Manfred Dollhausen	Odenthal
Dr. Hannes Frank	Detmold
Prof. Dr. Otto-D. Hennemann	Osterholz-Scharmbeck

REPORT 2016

FCIO – Austria

FCIO – Austria

The Austrian Association for Flooring adhesives was founded 2008, as a successor from the Austrian Adhesives Association, which was dissolved in 2007. We are a selfstanding part of the Austrian Chemical Industry Federation and the Federal Economic Chamber of Austria.

The Austrian Association of Flooring Adhesives has at the moment 16 members.

Mission statement and services

FCIO-Berufsgruppe Bauklebstoffe (Flooring Adhesive Group) with its 16 member companies, is a legal based association within the Chemical Industry Federation, which represents and defends the interests of the Flooring adhesive Industry in Austria. As an official body, we are mandated by law to give input to all legal and economic issues, which are relating to our business. We stay in permanent contact with our Authorities and our Trade Unions and have representatives in steering committees of various scientific institutions and in working groups of different Austrian Ministries as well as standardization bodies. We help our members to comply with all the demanding laws and regulations in the field of Safety, Health and Environment, especially REACH and CLP or VOC's by preparing various guidance papers and give them legal advice in labor relations.

Together with our clients we are also running education programs for apprentices and trainees.

Organisation and structure

We are part of the Austrian Chemical Industrie Federation FCIO and have our offices in Vienna, Austria

President: Bernhard Mucherl/Murexin
Email: b.mucherl@murexin.com

Director: Klaus Schaubmayr
Email: schaubmayr@fcio.at

REPORT 2016

FKS – Switzerland

FKS – Switzerland

Professional Association of the Swiss Adhesive and Sealant Industry

Mission Statement
The association supports its members with regard to adhesive and sealant production, in particular through:
- Representing the interests of the Swiss adhesive and sealant industry with regard to public authorities and associations, inclusive participation in legislative tasks
- Participation in professional expert panels to strengthen the cooperation with public authorities and national and international associations
- Statistics and base information about the Swiss adhesive and sealant market, which provides Swiss and European public authorities with the basic information needed to support their decision-making processes
- Technical clarification and expert reports to promote the customers confidence in the members of the association
- Regular exchanges of information and experience among the members in order to develop the quality of the products
- Organisation of specialised expert lectures

Market developments, directives and measures
Measures for environmental protection and safety in production, packaging, transport, application and disposal are based on the monitoring of market developments and the existing directives.

The measures help to ensure that the services provided always meet the highest requirements of the market.

Member of FEICA
(Association of the European Adhesive and Sealant Industry)
The association is a member of FEICA, which represents the interests of its national associations at the international level in relation to cooperation with international organisations.
FEICA regularly provides the national associations with information concerning developments in Europe.

Service Profile
- Sales statistics about turnover and volume in Switzerland from member companies
- National standards
- Technical clarification and expert reports
- Exchange of information and experience
- Information about REACH
- Annual meeting of members in spring and autumn
- Access to FKS information by the use of member log in on the internet

Organisation & Structure

- President
  Heinz Leibundgut
- Vice-President
  Marcel Leder-Maeder

Members:

Alfa Klebstoffe AG	Jowat Swiss AG
APM Technica AG	Kisling AG
Artimelt AG	merz+benteli ag
Astorit AG	nolax AG
Avery Dennison Materials Europe GmbH	Sika (Schweiz) AG
Collano Adhesives AG	Türmerleim AG
Emerell AG	Uzin Tyro AG
EMS-Griltech AG	Wakol Adhesa AG
H.B. Fuller Europe GmbH	ZHAW – Zurich University of Applied Sciences
Henkel & Cie. AG	

Contact Information

President	Vice-President	Secretariat
Heinz Leibundgut Uzin Tyro AG Ennetbürgerstrasse 47 6374 Buochs Phone: +41 (0)41 624 48 80 Fax: +41 (0)41 624 48 89 Email: heinz.leibundgut@ uzin-utz.com Email: info@fks.ch	Marcel Leder-Maeder Türmerleim AG Hauptstrasse 15 4102 Binningen Phone: +41 (0) 61 271 21 66 Fax: +41 (0) 61 271 21 74 Email: marcel.leder@ tuermerleim.ch	Fachverband Klebstoff- Industrie Schweiz Silvia Fasel Postfach 213 5401 Baden Phone: +41 (0) 56 221 51 00 Fax: +41 (0) 56 221 51 41 www.fks.ch

vereniging lijmen en kitten

REPORT 2016

VLK – Netherland

VLK – Netherland

VLK

Industrial Association

The Dutch Adhesives and Sealants Association (VLK) represents the adhesives and sealants industry in the Netherlands and protects a European level playing field for the technical and economic interests of its member companies.

The activities of the adhesives and sealants industry are primarily business to business. Adhesives and sealants are mainly used in the industrial and construction sectors. It is estimated that the VLK represents at least 80% of adhesive and sealant sales in the Netherlands.

For the member companies the VLK is:
• contact point for the government and government agencies such as appropriate authorities, labour inspectorate, users in the market, supply chain partners and other stakeholders;
• spokesman for the industry to achieve realistic legislation, on a national as well as European level;
• information source and helpdesk for terms and conditions concerning substances (like REACH, CLP, biocides etc.) and concerning construction regulation (like CPR and CE marking);
• protector of the image of adhesives and sealants, and of its industry;
• facilitator of knowledge sharing and networking.

The VLK is a member of FEICA (the Association of the European Adhesive & Sealant Industry) which promotes the interests of the sector at European level.

Services
The VLK provides the following services:

• Lobby
The VLK fights for practical terms and conditions for the development, production and sale of adhesives and sealants in the Netherlands. The association promotes the interests of the sector in its contacts with the government, umbrella organisations, buyer organisations and supply chain partners. Many new developments and initiatives originate from Brussels and for that reason the VLK is a member of FEICA, the Association of the European Adhesive & Sealant Industry.

• Complete network
The VLK brings members into contact with each other and with a large number of players within the adhesives and sealants industry. This goes beyond the own industry; the VLK encourages to exchange experiences with other partners in the value chain and with knowledge institutes.

• Sharing knowledge
The VLK filters relevant information and distributes it to members via the members' website or via the digital newsletter. Members can also consult up-to-date guides and publications from the members' website. The VLK has three sector groups and two technical committees where the members inform each other about new developments, for example with regard to new legislation, health and safety, the environment, and standards.

• Insight into market development
The floor sector group and the wall tile sector group provide information about developments in their market sectors. A reputable agency organizes the benchmark for the two sector groups and provides the participants each quarter a report on market developments.

• Helpdesk
Members can contact the VLK with questions relating to adhesives and sealants, for example about legislation affecting the sector. The VLK provides tailored advice or refers members to an organisation that is able to provide further help.

• Positive image and trust
Because of VLK's independent position, the industry association VLK is able to provide an objective view and general information externally. The VLK communicates about the (often invisible) contribution of adhesives and sealants to the comfort in our live.

Members
It is the members that make the VLK. The industrial association closely involves directors, managers and experts from its membership in the work that is carried out in the VLK. The contact details of all members can be found at www.vlk.nu/leden. The VLK does not get involved in the commercial activities of the individual companies.

Organisation
The General Members' Meeting elects the Board that manages the VLK. The VLK has the following Board members:
• Wybren de Zwart (Saba Dinxperlo BV) – Chairman
• Dirk Breeuwer (Forbo Eurocol BV)
• Rob de Kruijff (Sika Nederland BV)
• Gertjan van Dinther (Soudal BV)

The VLK has the following sector groups:
• Floor adhesives and levelling compounds
• Tile adhesives
• Sealants

The VLK has the following technical groups:
• Committee for Health, Safety and the Environment
• Technical Committee for Sealants

Office
Like paints and printing inks, adhesives and sealants are compound products. The VLK works closely with the Netherlands Association of Paint and Printing Ink Manufacturers (VVVF) with whom it shares an office. The VVVF has established a number of sector-wide themes in order to give structure and priority to its activities. Wherever possible the VLK joins these themed groups in order to gain synergy benefits from the collaboration with the VVVF.

Further information?
For further information please contact the VLK or visit www.vlk.nu.

VLK
Postbus 241
2260 AE Leidschendam, Netherlands
Tel.: + 3170 444 06 80
Email: info@vlk.nu
Website: www.vlk.nu

GEV

Association for the Control of Emissions
from Products for Flooring Installation,
Adhesives & Building Materials

EMICODE®

Protection from Indoor Air Contaminants

The need to protect people from indoor air contaminants resulted in the foundation of the Association for the Control of Emissions in Products for Flooring Installation, Adhesives and Construction Products (GEV) and the establishment of the EMICODE labelling system in February 1997. This initiative was supported by leading manufacturers of flooring installation products that are also members of the German Adhesives Association.

In the 1950s, the installers of parquet and other types of floor coverings were exposed to risks such as inhaling emissions, explosions of solvents and eczema caused by chromates. During this period, health and safety was regarded as less important than functionality and cost-effectiveness. In the decades that followed, the amounts of solvents used were gradually reduced until they were finally almost completely removed from the products in question. This change was supported by the hazardous substances information system (GISBAU) set up by BG Bau (the employers' liability insurance organisation for the construction industry), which was the first system to provide guidance relating to health and safety at work in this area. Technological developments during the 1990s made it possible to reduce significantly the quantities of volatile organic compounds used. It was against this background that the GEV was founded.

On 24 February 1997, well-known German manufacturers of flooring adhesives came together to provide processors and consumers with reliable guidance in the light of the large number of different measurement processes in use. They set up an objective qualification and labelling system called EMICODE which makes it possible to evaluate flooring installation materials and other building products on the basis of their emissions. At the same time, it gives manufacturers a strong incentive to develop their products further.

The EMICODE classes are based on a precisely defined test chamber examination and demanding classification criteria. Adhesives, levelling compounds, precoats, substrates, sealants, fast-drying screeds and other building products which are marked with the GEV label EMICODE EC 1 as "very low in emissions" produce the lowest possible level of indoor air contaminants. Unlike other systems, with EMICODE the manufacturers themselves are responsible for labelling, while GEV has samples taken of products on the market by independent institutes for checking purposes. Another difference is that GEV does not allow for any compromises on quality. Technically questionable environmental criteria are not permitted for reasons of sustainability.

This voluntary initiative is a systematic continuation of the efforts to protect the health of processors and consumers. EMICODE provides contracting authorities, architects, planners, tradespeople, building owners and end consumers with transparent and objective guidelines for selecting low-emission flooring installation products. Videos in 11 languages, brochures, tender document templates, technical documents and the organisation's regulations are available on the website www.emicode.com.

As a result of pressure from the market, the scope of EMICODE, which had originally focused only on flooring installation products, was expanded to cover a wide range of construction chemicals in almost all areas of interior finishing. By extending EMICODE to include products such as joint sealants, parquet varnishes, polyurethane foams and window films, GEV was responding to market requirements for the classification of other products and also to other calls from the market and from industry to differentiate between products on the basis of their environmental impact. For products which are subject to labelling requirements and which, although their emission levels are low, require measures to be taken to ensure safe processing, GEV has introduced the "R" suffix ("regulated") in the logo.

For 20 years, EMICODE® has been making an important contribution to creating healthy living environments and promoting sustainable construction practices. It is also filling an important gap. The protective measures for flooring installation materials imposed by the system not only cover processors but also consumers and the environment. Finally, guidelines have been introduced for sustainable, low-emission products that contribute to a healthy living environment. The focus is on short-term and long-term emissions, together with very low levels of volatile organic compounds.

Over the last 20 years, EMICODE® has become one of the most important consumer and environmental protection schemes in the industry and has now spread outside Europe. Its emissions limits are currently the toughest standards on the market. The label provides an objective benchmark for the emissions of construction products that has been tested by independent experts and analysis laboratories. In contrast to other health and environmental compatibility certification systems, the scheme involves taking random samples to ensure that certified products comply with the requirements.

GEV currently has 124 member companies, around half of which are based outside Germany. More than 5,000 products have been certified using the process and this figure is constantly growing.

The members of GEV, the licensing and monitoring body, include experts from different sectors of the construction industry. Everyone involved in the value chain, from suppliers of raw materials to product manufacturers, is working towards the same objective. EMICODE® and emission levels are now the key sustainability factors for organisations such as the Institute for Construction and the Environment (IBU), the German Sustainable Building Council (DGNB) and the US Green Building Council (LEED).

Chairman of the Executive Board   Stefan Neuberger, Pallmann
Chairman of the Technical Advisory Board Jürgen Gehring, Bostik GmbH
Managing Director of GEV   Klaus Winkels, Attorney-at-law

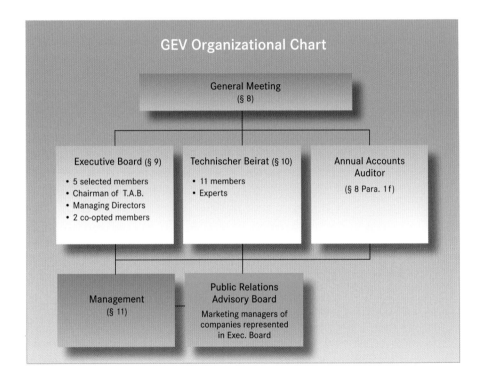

Gemeinschaft Emissionskontrollierte Verlegewerkstoffe,
Klebstoffe und Bauprodukte (GEV)
RWI-Haus
Völklinger Straße 4
D-40219 Düsseldorf
Phone +49 (0) 2 11-6 79 31-20
Fax   +49 (0) 2 11-6 79 31-33
Email: info@emicode.com
www.emicode.com

Association of the European
Adhesive and Sealant Industry

FEICA – Fédération Européenne des Industries de Colles et Adhésifs (Association of the European Adhesive and Sealant Industry) – has represented the interests of Europe's adhesives industry since 1972. As the umbrella organisation for 14 national adhesives associations in Europe, it also represents the common interests of some individual enterprises, as well as primarily internationally operating adhesive manufacturers and producers of polyurethane foams.

The office of FEICA was operated until the end of 2006 together with the German Adhesives Association with headquarters in Düsseldorf, Germany. Since January 2007, FEICA's headquarters have been located in Brussels.

FEICA's Objectives

With the support of its members, FEICA advocates the common interests of our industry throughout Europe. Moreover, it is also responsible for representing the interests of its members at institutions of the European Union.

External Relations

To obtain information about projects and decrees of European institutions as quickly as possible (European Parliament, European Council, European Commission, Directorate General), FEICA maintains close contact to **CEFIC** (European Chemical Industry Council) and other European associations, of which many from the DUCC group (**D**ownstream **U**ser of **C**hemicals **C**o-ordination) have joined.

Service for IVK Members

As the largest member of FEICA, IVK represents the interests of German adhesive manufacturers in the Executive Board, in the European Technical Board and in various other specialised committees of the European association. This combination assures the members of the German Adhesives Association a corresponding, free-of-charge service as well as all necessary information and regular contact on a European level.

Contact Address

FEICA – Avenue E. van Nieuwenhuyse, BE 1160 Brussels www.feica.eu

Relevant Laws and Regulations
for Adhesives

Hazardous Material Law

- **REACH Regulation** – Regulation (EC) No 1907/2006 of the European Parliament and of the Council of 18 December 2006 concerning the Registration, Evaluation, Authorization and Restriction of Chemicals (REACH), establishing a European Chemicals Agency, amending Directive 1999/45/EC and repealing Council Regulation (EEC) No 793/93 and Commission Regulation (EC) No 1488/94 as well as Council Directive 76/769/EEC and Commission Directives 91/155/EEC, 93/67/EEC, 93/105/EEC and 2000/21/EC
- **CLP Regulation** – Regulation (EC) No 1272/2008 of the European Parliament and of the Council of 16 December 2008 on classification, labelling and packaging of substances and mixtures, amending and repealing Directives 67/548/EEC and 199/45/EC and amending Regulation (EC) No 1907/2006
- **Fees Regulation REACH** – Commission Regulation (EC) No 340/2008 of 16 April 2008 on the fees and charges payable to the European Chemicals Agency pursuant to Regulation (EC) No 1907/2006 of the European Parliament and of the Council on the Registration, Evaluation, Authorization and Restriction of Chemicals (REACH)
- **Fees Regulation CLP** – Commission Regulation (EU) No 44/2010 of 21 May 2010 on the fees payable to the European Chemicals Agency pursuant to Regulation (EC) No 1272/2008 of the European Parliament and of the Council on classification, labelling and packaging of substances and mixtures
- **Chemicals test method Regulation** – Council Regulation (EC) No 440/2008 of 30 May 2008 laying down test methods pursuant to Regulation (EC) No 1907/2006 of the European Parliament and of the Council on the Registration, Evaluation, Authorization and Restriction of Chemicals (REACH)
- **Biocidal Products Regulation** – Regulation (EU) No 528/2012 of the European Parliament and of the Council of 22 May 2012 concerning the making available on the market and use of biocidal products
- **PIC Regulation** – Regulation (EC) No 649/2012 of the European Parliament and of the Council concerning the export and import of hazardous chemicals (The new text was adopted on 4 July 2012 and will be applicable from 1 March 2014)
- **POP Regulation** – Regulation (EC) No 850/2004 of the European Parliament and of the Council of 29 April 2004 on persistent organic pollutants and amending Directive 79/117/EEC

Waste Legislation

- **Waste Framework Directive** – Directive 2008/98/EC of the European Parliament and of the Council of 19 November 2008 on waste and repealing certain Directives
- **List of wastes** – Commission Decision 2000/532/EC of 3 May 2000 replacing Decision 94/3/EC establishing a list of wastes pursuant to Article 1(a) of Council Directive 75/442/EEC on waste and Council Decision 94/904/EC establishing a list of hazardous waste pursuant to Article 1(4) of Council Directive 91/689/EEC on hazardous waste
- **Packaging Waste Directive** – European Parliament and Council directive 94/62/EC of 20 December 1994 on packaging and packaging waste

Immission protection

- **Directive on Ambient Air Quality** - Directive 2008/50/EC of the European Parliament and of the Council of 21 May 2008 on ambient air quality an cleaner air for Europe
- **Directive on industrial emissions** – Directive 2010/75/EU of the European Parliament and of the Council of 24 November 2010 on industrial emissions (integrated pollution prevention and control)
- **VOC emissions, paints and varnishes** – Directive 2004/42/CE of the European Parliament and of the Council of 21 April 2004 on the limitation of emissions of volatile organic compounds due to the use of organic solvents in certain paints and varnishes and vehicle refinishing products and amending Directive 1999/13/EC
- **Emissions Trading Directive** – Directive 2003/87/EC of the European Parliament and of the Council of 13 October 2003 establishing a scheme for greenhouse gas emission allowance trading within the Community and amending Council Directive 96/61/EC
- **PRTR Regulation** – Regulation (EC) No 166/2006 of the European Parliament and of the Council of 18 January 2006 concerning the establishment of a European Pollutant Release and Transfer Register and amending Council Directives 91/689/EEC and 96/61/EC
- **F-Gases Regulation** – Regulation (EU) No 517/2014 of the European Parliament and of the Council of 16 April 2014 on fluorinated greenhouse gases and repealing Regulation (EC) No 842/2006

Water Legislation

- **Water Framework Directive** – Directive 2000/60/EC of the European Parliament and of the Council of 23 October 2000 establishing a framework for Community action in the field of water policy
- **Dangerous Substances Directive** – Directive 2006/11/EC of the European Parliament and of the Council of 15 February 2006 on pollution caused by certain dangerous substances discharged into the aquatic environment of the Community

Hazardous Materials Transportation Law

- Council Directive 96/49/EC of 23 July 1996 on the harmonization of the laws of the Member States with regard to the transport of dangerous goods by rail
- Council Directive 94/55/EC of 21 November 1994 on the harmonization of the laws of the Member States with regard to the transport of dangerous goods by road
- Council Directive 95/50/EC of 6 October 1995 on uniform procedures for checks on the transport of dangerous goods by road
- Directive 98/91/EC of the European Parliament and of the Council of 14 December 1998 relating to motor vehicles and their trailers intended for the transport of dangerous goods by road and amending Directive 70/156/EEC relating to the type approval of motor vehicles and their trailers
- Regulation (EC) No 2099/2002 of the European Parliament and of the Council of 5 November 2002 establishing a Committee on Safe Seas and the Prevention of Pollution from Ships (COSS) and amending the Regulations on maritime safety and the prevention of pollution from ships
- Council Directive 96/35/EC of 3 June 1996 on the appointment and vocational qualification of safety advisers for the transport of dangerous goods by road, rail and inland waterways
- Directive 2008/68/EC of the European Parliament and of the Council of 24 September 2008 on the inland transport of dangerous goods

International Treaties
GHS – Globally Harmonized System of Classification and Labelling of Chemicals

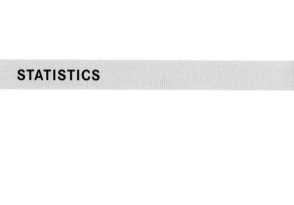

STATISTICS

German Adhesives Industry
Production of Adhesives Systems 2016

➢ Adhesives	916.000 t
➢ Sealants	174.000 t
➢ cement-based Systems	410.000 t
Total	1.500.000 t
➢ Tapes	1.054 Mio. m²

Source: IVK

© Industrieverband Klebstoffe e. V., Düsseldorf

German Adhesives Industry
Adhesives Systems 2016

	2011	2012	2013	2014	2015	2016	
Adhesives	1,66	1,71	1,76	1,81	1,84	1,85	
Sealants	0,56	0,58	0,63	0,65	0,67	0,68	
cement-based Systems	0,36	0,37	0,38	0,39	0,40	0,41	
Tapes	0,72	0,74	0,76	0,78	0,80	0,81	
Total	3,30	3,40	3,53	3,63	3,71	3,75	in bn. €

Source: IVK

© Industrieverband Klebstoffe e. V., Düsseldorf

Development of selected Industries in Germany

	% (of 2017)	2015	Prognosis 2016	Prognosis 2017
Manufacturing	**100**	**2,6**	**3,0**	**3,3**
Transportation	2.9	1,6	2,4	2,7
Food, Beverage & Tobacco	13.2	2,1	3,0	2,9
Paper / Printed Matters	3.1	1,2	1,0	2,0
Metals & Metal Products	13.8	1,6	1,4	2,6
Plant & Machinery	7.5	0,0	1,7	3,3
Electrical & Optical Equipment	5.0	3.5	5.1	5.2
Chemistry	8.2	3.6	3.7	3.1
Wood (without Furniture)	1.4	2,9	4,2	4,2
Construction Industry	**–**	**2,6**	**2,5**	**3,0**

Source: Stat. Bundesamt, Prognosis FERI; May 2015

© Industrieverband Klebstoffe e. V., Düsseldorf

German Adhesives Industry
- Global Position -

World Market*
ca. 60 Mrd. € Turnover Domestic Market
1,9 Mrd. € Exports
1,6 Mrd. € Turnover Subsidiaries outside Germany
7,7 Mrd. € Total
11,2 Mrd. € World Market Share

19 %

* Matter to market definition & exchange rate effects

Source: IVK

© Industrieverband Klebstoffe e. V., Düsseldorf

Development of Raw Material Price Index

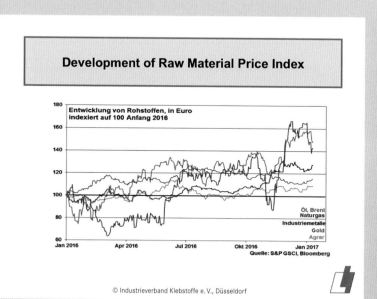

© Industrieverband Klebstoffe e. V., Düsseldorf

Adhesives Industry 2017

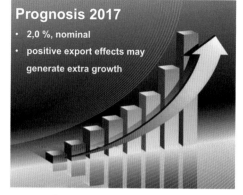

Prognosis 2017

- 2,0 %, nominal
- positive export effects may generate extra growth

© Industrieverband Klebstoffe e. V., Düsseldorf

STANDARDS

Adhesives standards

Early in the 1950s, adhesives standardisation started in Germany at a national level. This initiative continued with the issuing of a terminology standard and other standards for various fields (wood, shoes, flooring adhesives). DIN standards issued during that period by the German Institute for Standardisation (DIN) still apply on a domestic level, unless they have been repealed or replaced by European standards. On 23 January 2013, DIN formed the new committee "Adhesives; Test Procedures and Requirements", which will in future function as an umbrella committee for the DIN standards committees that deal with adhesives. The reason is that most of the standards required are in the meantime available in the form of DIN EN standards, thus making national standards committees dormant. This had the consequence that, in those cases in which a revision was decided on a European level, no active mirror committee and therefore no possibilities for participation were available. The formation of the committee "Adhesives; Test Procedures and Requirements" thus represents a milestone in the reorganisation of adhesives committees at DIN. It is now possible to spin off new subordinate committees relatively quickly and therefore to react more quickly to new requirements than was previously the case. Therefore and due to the integration of other DIN committees dealing with adhesives, the NMP (Standards Committee Materials Testing) sees itself as being well equipped for the future.

In 1961, the "COMITÉ EUROPÉEN DE NORMALISATION" (European Committee for Standardization, CEN) was founded. Thereafter, CEN established various technical committees (Technical Committee 52 "Safety of toys", Technical Committee 67 "Adhesives for tiles", Technical Committee 193 "Adhesives", Technical Committee 261 "Packaging" and Technical Committee 264 "Indoor air quality") to establish and revise standards in the field of adhesives. Thus for the first time the opportunity was created to develop a broad set of standards applicable throughout Europe, and which was specifically aimed at adhesives manufacturers and consumers. This brought with it the chance for them to better represent their own economic and technical interests. Agreements which CEN signed with the "INTERNATIONAL ORGANIZATION FOR STANDARDIZATION" (ISO) meant that the global ISO standards were accepted as European standards and the two organisations went on to jointly work out and issue additional standards.

European standards are generally publicised in the three official languages of the European Union, i. e., German, English and French, and conform with one another from a technical perspective. European standardisation supports trade across the EU by reducing trade barriers and is part of the framework of efforts to achieve technical harmonisation. European standardisation is supported by the German Adhesives Association. Unlike other standards, e. g., international ISO standards, European standards (EN) are binding at a European level. Both the European Court of Justice and national courts within the EU are obliged to base their jurisdiction on European standards.

Specialist testing of adhesives and bonded joints is an important prerequisite for the success of the bonding process. The German Adhesives Association (IVK) has developed a new online tool, which is now available at www.klebstoffe.com, to provide users with support in this area.

The tool is a database that for the first time summarises all the relevant standards, directives and other important guidelines at a glance. The list currently consists of more than 750 individual documents. It includes not only the relevant German, European and international standards (DIN, EN, ISO), but also standards that were not specifically created for use in the field of bonding, but have become well established in this area.

All the documents are categorised on the basis of the type of adhesive, the application and the contents of the document. Practical search functions allow users to find the subject areas that are relevant to them. A brief description of the contents of each document helps users to determine whether it answers their questions or meets their specific testing requirements. If the document is available free of charge, there is a link to the original version. If payment is required, a link is provided to the website where the document can be purchased.

The standards database is kept constantly updated by IVK, in collaboration with DIN Software GmbH and the consultancy company KLEBTECHNIK Dr. Hartwig Lohse e. K. This ensures that users can always identify which testing methods for adhesives and bonded joints meet their needs. http://www.klebstoffe.com/start-normentabelle.html

SOURCES

- Raw Materials
- Adhesives by Types
- Sealants
- Adhesives by Key Market Segments
- Equipment
- Technical Consultancy
- Contract Manufacturing and Filling Services
- Research and Development

Raw Materials

3M
Alberdingk Boley
ARLANXEO
Avebe Adhesives
BASF
Biesterfeld Spezialchemie
Bodo Möller
Brenntag
BYK
Celanese
Chemische Fabrik Budenheim
CHT R. Breitlich
CnP Polymer
Coim
Collall
Collano
Covestro
CTA GmbH
DKSH GmbH
EMS-Chemie
Evonik
ExxonMobil Chemical
IMCD
KANEKA Belgium
Keyser & Mackay
Krahn Chemie
LANXESS
NAGASE
Nordmann, Rassmann
Michelman
Münzing
Omya Hamburg GmbH
ORGANIK KIMYA
Poly-Chem
Rütgers
Schill+Seilacher
Schülke & Mayr GmbH
Sonderhoff
Synthomer Deutschland GmbH
Synthopol Chemie
Ter Hell
versalis S.p.A.
Wacker Chemie

Willers, Engel
Worlée-Chemie

Adhesives by Types

Hot-melt Adhesives

3M
ALFA Klebstoffe AG
ARDEX
Artimelt
Beardow Adams
Biesterfeld Spezialchemie
Bodo Möller
Bostik
BÜHNEN
BYLA
CHT R. Beitlich
Collano
Dow
Drei Bond
Eluid Adhesive
EMS-Chemie
EUKALIN
Evonik
ExxonMobil Chemical
Follmann
Forbo Eurocol
H.B. Fuller
GLUDAN
Fritz Häcker
Henkel
ISP
Jowat
Kleiberit
Kömmerling
NAGASE
Paramelt
Planatol
Poly-Chem
PRHO-CHEM
Rampf
Rhenocoll
Ruderer Klebtechnik
SABA Dinxperlo

Siema
Sika Automotive
Sika Deutschland
Tremco illbruck
TSRC (Lux.) Corporation
Türmerleim
Unitech
versalis S.p.A.
VITO Irmen
Weiss Chemie + Technik
Zelu

Reactive Adhesives
3M
Adtracon
ARDEX
Berger-Seidle Siegeltechnik
Biesterfeld Spezialchemie
BLUFIXX
Bona
Bodo Möller
Bolton Adhesives
Bostik
BÜHNEN
BYLA
Chemetall
COIM Deutschland
Collano
Cyberbond
DEKA
DELO
Den Braven Benelux BV
Dow
Drei Bond
Dymax Europe
fischer
Forbo Eurocol
Gößl + Pfaff
H.B. Fuller
Henkel
Jowat
Kiesel Bauchemie
Kisling Deutschland GmbH
Kleiberit

Kömmerling
LORD
LOOP
LUGATO CHEMIE
merz+benteli
NAGASE
Otto-Chemie
Panacol-Elosol
Paramelt
PCI
Planatol
Rampf
Ramsauer GmbH
Ruderer Klebtechnik
SABA Dinxperlo
SCA Schucker
Schlüter
Schönox
Schomburg
Siema
Sika Automotive
Sika Deutschland
Sonderhoff
STAUF
Stockmeier
Synthopol Chemie
Tremco illbruck
Uzin Tyro
Uzin Utz
Vinavil
Wakol
Weicon
Weiss Chemie + Technik
ZELU CHEMIE

Dispersion Adhesives
3M
ALFA Klebstoffe AG
ATP adhesives systems AG
Berger-Seidle Siegeltechnik
Biesterfeld Spezialchemie
Bodo Möller
Bona
Bostik

BÜHNEN
Celanese
CHT R. Beitlich
Coim
Collall
Collano
CTA GmbH
DEKA
Den Braven Benelux BV
Drei Bond
Eluid Adhesive
EUKALIN
fischer
Follmann
Forbo Eurocol
H.B. Fuller
GLUDAN
Fritz Häcker
Henkel
IMC
ISP
Jowat
Kiesel Bauchemie
Klebstoffwerk COLLODIN
Kleiberit
Kömmerling
LORD
LUGATO CHEMIE
Michelman
Murexin
ORGANIK KIMYA
Paramelt
PCI
Planatol
PRHO-CHEM
Ramsauer GmbH
Renia-Gesellschaft
Rhenocoll
Ruderer Klebtechnik
SCA Schucker
Schlüter
Schönox
Schomburg
Sika Automotive
Siema

Sopro Bauchemie
STAUF
Synthopol Chemie
Tremco illbruck
Türmerleim
UHU
VITO Irmen
Wakol
Weiss Chemie & Technik
Wulff
ZELU CHEMIE

Vegetable Adhesives,
Dextrin and Starch Adhesives
Beardow Adams
Biesterfeld Spezialchemie
Bodo Möller
Collall
Eluid
EUKALIN
H.B. Fuller
Henkel
Paramelt
Planatol
PRHO-CHEM
Ruderer Klebtechnik
Schönox
Siema
Türmerleim

Animal Glue
H.B. Fuller
Henkel
PRHO-CHEM

Solvent-based Adhesives
3M
Adtracon
Berger-Seidle Siegeltechnik
Biesterfeld Spezialchemie
Bodo Möller
Bona
Bolton Adhesives
Bostik
CHT R. Beitlich

COIM Deutschland
Collall
CTA GmbH
DEKA
Den Braven Benelux BV
Fermit
fischer
Forbo Eurocol
H.B. Fuller
IMCD
Jowat
Kiesel Bauchemie
Kleiberit
Kömmerling
LANXESS
LORD
NAGASE
Otto-Chemie
Paramelt
Planatol
Poly-Chem
Ramsauer GmbH
Renia-Gesellschaft
Rhenocoll
Ruderer Klebtechnik
SABA Dinxperlo
SCA Schucker
Schönox
Siema
Sika Automotive
STAUF
Synthopol Chemie
Tremco illbruck
TSRC (Lux.) Corporation
UHU
versalis S.p.A.
VITO Irmen
Wakol
Weiss Chemie + Technik
ZELU CHEMIE

Pressure-Sensitive Adhesives
3M
ALFA Klebstoffe AG
ATP adhesives systems AG

Beardow Adams
Biesterfeld Spezialchemie
Bostik
BÜHNEN
Collano
CTA GmbH
DEKA
Dymax Europe
Eluid Adhesive
EUKALIN
H.B. Fuller
GLUDAN
Fritz Häcker
Henkel
IMCD
Kleiberit
LANXESS
NAGASE
ORGANIK KIMYA
Paramelt
Planatol
Poly-Chem
PRHO-CHEM
Rhenocoll
Ruderer Klebtechnik

Sealants

ARDEX
Berger-Seidle Siegeltechnik
Bodo Möller
Bolton Adhesives
Bostik
Botament
CTA GmbH
Drei Bond
ExxonMobil
Den Braven Benelux BV
EMS-Chemie
Fermit
fischer
Henkel
merz+benteli
Murexin
NAGASE

ORGANIK KIMYA
OTTO-Chemie
Paramelt
PCI
Rampf
Ramsauer GmbH
Ruderer Klebtechnik
SCA Schucker
Schomburg
Sonderhoff
Stockmeier
Synthopol Chemie
Tremco illbruck
UHU
Wulff

Adhesives by
Key Market Segments
Self Adhesive Tapes
3M
Alberdingk Boley
artimelt
ATP adhesives systems AG
Avebe Adhesives
Bodo Möller
BYK
certoplast Technische Klebebänder
CNP-Polymer
Coroplast
DKSH GmbH
Eluid Adhesive
Fritz Häcker
IMCD
LANXESS
Lohmann
Planatol
Schlüter
Synthopol Chemie
Tesa

Paper/Packaging
3M
Adtracon
Alberdingk Boley

ALFA Klebstoffe AG
artimelt
ATP adhesives systems AG
Avebe Adhesives
Beardow Adams
Biesterfeld Spezialchemie
Bodo Möller
Bostik
Brenntag
BÜHNEN
BYK
Celanese
certoplast Technische Klebebänder
CNP-Polymer
COIM Deutschland
Collano
Coroplast
CTA GmbH
DEKA
DKSH GmbH
Eluid Adhesive
EMS-Chemie
EUKALIN
Evonik
ExxonMobil
Follmann
Forbo Eurocol
H.B. Fuller
GLUDAN
Fritz Häcker
Henkel
IMCD
Jowat
LANXESS
Lohmann
Michelman
MÜNZING
NAGASE
Nordmann, Rassmann
Nynas
Omya Hamburg GmbH
ORGANIK KIMYA
Paramelt
Planatol
Poly-Chem

PRHO-CHEM
Rhenocoll
Ruderer Klebtechnik
Schülke & Mayr GmbH
Siema
Sonderhoff
Synthopol Chemie
tesa
TSRC (Lux.) Corporation
Türmerleim
UHU
versalis S.p.A.
Wakol
Weicon
Weiss Chemie + Technik

Bookbinding/
Graphics Industry
3M
ALFA Klebstoffe AG
ATP adhesives systems AG
Biesterfeld Spezialchemie
Bodo Möller
Brenntag
BÜHNEN
BYK
Celanese
CNP -Polymer
Coim
Collall
DKSH GmbH
Eluid Adhesive
EUKALIN
Evonik
ExxonMobil
H. B. Fuller
Fritz Häcker
Henkel
IMCD
Jowat
LANXESS
Lohmann
Michelman
MÜNZING
Nordmann, Rassmann

Omya Hamburg GmbH
ORGANIK KIMYA
Planatol
PRHO-CHEM
Siema
Sika Automotive
tesa
TSRC (Lux.) Corporation
Türmerleim
UHU
versalis S.p.A.
Vinavil

Wood/Furniture industry
3M
Adtracon
ALFA Klebstoffe AG
ATP adhesives systems AG
Berger-Seidle Siegeltechnik
Biesterfeld Spezialchemie
BLUFIXX
Bodo Möller
Bolton Adhesives
Bostik
Brenntag
BÜHNEN
BYK
BYLA
Celanese
Chemische Fabrik Budenheim
CNP-Polymer
Collall
Collano
Coroplast
CTA GmbH
Cyberbond
DEKA
Den Braven Benelux BV
DKSH GmbH
Eluid Adhesive
EMS-Chemie
Evonik
ExxonMobil
fischer
Follmann

Gößl + Pfaff
H.B. Fuller
Henkel
Jowat
KANEKA Belgium
Kisling Deutschland GmbH
Kleiberit
Kömmerling
LANXESS
Lohmann
merz+benteli
Michelman
MÜNZING
Nordmann
Omya Hamburg GmbH
ORGANIK KIMYA
Otto-Chemie
Panacol-Elosol
Rampf
Ramsauer GmbH
Rhenocoll
Ruderer Klebtechnik
SABA Dinxperlo
Siema
Sika Automotive
STAUF
Stockmeier
Synthopol Chemie
tesa
Tremco illbruck
TSRC (Lux.) Corporation
Türmerleim
versalis S.p.A.
Vinavil
VITO Irmen
Wakol
Weicon
Weiss Chemie + Technik
ZELU CHEMIE

Building and Construction Industry
including Floors, Walls and Ceilings
ARDEX
artimelt
Berger-Seidle Siegeltechnik

Biesterfeld Spezialchemie
BLUFIXX
Bodo Möller
Bona
Bostik
Botament
Brenntag
BÜHNEN
BYLA
BYK
Celanese
certoplast Technische Klebebänder
Chemische Fabrik Budenheim
CnP Polymer
Collano
Coroplast
CTA GmbH
DEKA
DELO
Den Braven Benelux BV
DKSH GmbH
Emerell
EMS-Chemie
Evonik
ExxonMobil
Fermit
fischer
Forbo Eurocol
Gößl + Pfaff
H. B. Fuller
GLUDAN
Henkel
IMCD
Kiesel Bauchemie
Kleiberit
Kömmerling
Lohmann
LUGATO CHEMIE
Mapei
Murexin
Michelman
MÜNZING
Nordmann, Rassmann
ORGANIK KIMYA
Otto-Chemie

Paramelt
Planatol
PCI
Poly-Chem
Rampf
Ramsauer GmbH
Rhenocoll
Schlüter
Schönox
Schomburg
Schülke & Mayr GmbH
Sika Automotive
Sika Deutschland
Sopro Bauchemie
STAUF
Synthopol Chemie
tesa
Tremco illbruck
TSRC (Lux.) Corporation
Uzin Tyro
Uzin Utz
Vinavil
Wakol
Weicon
Weiss Chemie + Technik
Wulff

Car and Aircraft Industries
ALFA Klebstoffe AG
APM Technica
ATP adhesives systems AG
Beardow Adams
BLUFIXX
Bodo Möller
Brenntag
BÜHNEN
BYLA
Celanese
certoplast Technische Klebebänder
Chemetall
Chemische Fabrik Budenheim
CHT R. Beitlich
CNP-Polymer
Coroplast
Cyberbond

DEKA
DELO
Den Braven Benelux BV
Dow
Drei Bond
Dymax Europe
Emerell
EMS-Chemie
Evonik
ExxonMobil
Gößl + Pfaff
H.B. Fuller
Henkel
Hönle
ISP
Kleiberit
Kömmerling
Lohmann
LORD
Michelman
MÜNZING
NAGASE
Nordmann, Rassmann
Otto-Chemie
Panacol-Elosol
Planatol
Polytec
Rampf
Ramsauer GmbH
Ruderer Klebtechnik
SCA Schucker
Schülke & Mayr GmbH
Sika Automotive
Sika Deutschland
Sonderhoff
Synthopol Chemie
Tremco illbruck
tesa
TSRC (Lux.) Corporation
Unitech
VITO Irmen
Wakol
Weicon
Weiss Chemie + Technik
ZELU CHEMIE

Electronics
APM Technica
ATP adhesives systems AG
BLUFIXX
Bodo Möller
Brenntag
BÜHNEN
BYLA
certoplast Technische Klebebänder
Chemetall
CHT R. Beitlich
Collano
Coroplast
CTA GmbH
Cyberbond
DELO
DKSH GmbH
Drei Bond
Dymax Europe
Emerell
EMS-Chemie
Evonik
ExxonMobil
Gößl + Pfaff
H.B. Fuller
Henkel
Hönle
KANEKA Belgium
Kisling Deutschland GmbH
Kömmerling
Lohmann
LORD
Michelman
MÜNZING
NAGASE
Nordmann, Rassmann
Otto-Chemie
Panacol-Elosol
Polytec
Rampf
Ruderer Klebtechnik
SCA Schucker
Siema
Sika Automotive
Sika Deutschland

tesa
Tremco illbruck
Unitech
UHU
Weicon
Weiss Chemie + Technik

Sanitary Industry
APM Technica
BLUFIXX
H.B. Fuller
GLUDAN
Henkel
ISP
Jowat
Kömmerling
LANXESS
Lohmann
Nordmann, Rassmann
Nynas
Prho-Chem
Schülke & Mayr GmbH
Sika Automotive
Türmerleim
Vito Irmen

Assembly, General Industry
Biesterfeld Spezialchemie
Bodo Möller
BÜHNEN
BYLA
certoplast Technische Klebebänder
Chemetall
CHT R. Beitlich
Coroplast
Cyberbond
DEKA
DELO
Dow
Drei Bond
ExxonMobil
Gößl + Pfaff
Henkel
KANEKA Belgium
Kleiberit

Kömmerling
Lohmann
Novamelt
Otto-Chemie
Panacol-Elosol
Paramelt
Renia-Gesellschaft
Ruderer Klebtechnik
SABA Dinxperlo
Schomburg
Schülke & Mayr GmbH
Sonderhoff
Synthopol Chemie
tesa
Weicon

Textile Industry
Adtracon
Biesterfeld Spezialchemie
Bodo Möller
Bostik
Brenntag
BÜHNEN
Chemische Fabrik Budenheim
CHT R. Beitlich
CNP-Polymer
Collano
DEKA
Emerell
EMS-Chemie
EUKALIN
Evonik
ExxonMobil
H.B. Fuller
Henkel
Jowat
Kleiberit
LANXESS
Michelman
MÜNZING
Nordmann, Rassmann
Novamelt
Nynas
Omya Hamburg GmbH
SABA Dinxperlo

Schülke & Mayr GmbH
Sika Automotive
Synthopol Chemie
tesa
Vito Irmen
Wakol
Wulff
Zelu

Self Adhesive Tapes, Labels
artimelt
Biesterfeld Spezialchemie
Bodo Möller
Bostik
Brenntag
CNP-Polymer
Coim
Collano
EMS-Chemie
EUKALIN
ExxonMobil
H.B. Fuller
Henkel
IMCD
Jowat
KANEKA Belgium
LANXESS
Michelman
MÜNZING
Nordmann, Rassmann
Nynas
ORGANIK KIMYA
Paramelt
Novamelt
Planatol
PRHO-CHEM
Schülke & Mayr GmbH
Stauf
Sika Automotive
Synthopol Chemie
TSRC (Lux.) Corporation
Türmerleim
versalis S.p.A.
Vito Irmen

Household, Hobby,
Offices, Stationery
BLUFIXX
Bodo Möller
Bolton Adhesives
certoplast Technische Klebebänder
CNP-Polymer
Collall
Coroplast
CTA GmbH
Cyberbond
Den Braven Benelux BV
EMS-Chemie
ExxonMobil
Fermit
fischer
GLUDAN
Henkel
KANEKA Belgium
LUGATO Chemie
Michelman
Nordmann, Rassmann
Nynas
Omya Hamburg GmbH
Panacol-Elosol
Rampf
Ramsauer GmbH
Renia-Gesellschaft
Rhenocoll
Schülke & Mayr GmbH
tesa
Tremco illbruck
TSRC (Lux.) Corporation
UHU
versalis S.p.A.
Weicon
Weiss Chemie + Technik

Footwear & Leather Industry
Adtracon
BÜHNEN
Cyberbond
H.B. Fuller
Henkel
Kömmerling

Renia-Gesellschaft
Ruderer Klebtechnik
Sika Automotive
Wakol
Zelu

Equipment

for Adhesive Handling, Mixing, Dosing
and Application
BÜHNEN
Drei Bond
Gößl + Pfaff
Graco
Hardo
Dr. Hönle
Innotech
IST Metz
Nordson
Reinhardt Technik (Wagner)
Robatech
SCA Schucker
Scheugenpflug
Sonderhoff
Sulzer Mixpac
TechconSystems
t-s-i.de Misch- und Dosiertechnik
Viscotec
Walther

Technical Consultancy

ChemQuest Europe INC.
Hinterwaldner Consulting
Klebtechnik Dr. Hartwig Lohse

Contract Manufacturing and Filling Services

LOOP

Research and Development

IFAM
ZHAW